Francois Gordon

Francois Gordon

Will Purdom 波爾登

Agitator, plant-hunter, forester

MMXXI

ISBN 978-1-910877-37-1

© Royal Botanic Garden Edinburgh, 2021.
Published by the Royal Botanic Garden Edinburgh
20A Inverleith Row, Edinburgh, EH3 5LR
www.rbge.org.uk

Francois Gordon has asserted his right under the Copyright, Design and Patents Act 1988 to be identified as the author of this work.
All rights reserved. No part of this publication may be reproduced, stored in a retrieval system or transmitted, in any form, or by any means, electronic, mechanical, photocopying, recording or otherwise, without the prior written consent of the copyright holders and publishers.

Proceeds from sales of this book will be used to support the work of RBGE.

The Royal Botanic Garden Edinburgh is a Non Departmental Public Body (NDPB) sponsored and supported through Grant-in-Aid by the Scottish Government's Environment and Forestry Directorate (ENFOR).

The Royal Botanic Garden Edinburgh is a Charity registered in Scotland (number SC007983).

Edited by Donna Cole, Royal Botanic Garden Edinburgh
Layout by Caroline Muir, Royal Botanic Garden Edinburgh
Indexed by Marie-Pierre Evans
Maps by Jeff Edwards
Printed by McAllister Litho Glasgow Limited

1056915
Printed on Carbon Captured paper

Introduction

I first heard of Will Purdom in June 1986, when I read a *Country Life* article about him written by the late Shirley Heriz-Smith. I knew just enough about the history of plant-hunting to know that Purdom's collecting in China had received very little attention from horticultural historians, and just enough about the plants he had introduced to Britain, as listed by Dr. Heriz-Smith, to understand that he deserved better. And the photograph illustrating the article – at that time believed to be the only surviving picture of Purdom – was intriguing.

At first glance, this portrait of a granite-jawed Purdom disguised as a Chinese muleteer is straight out of an Edwardian issue of the *Boy's Own Paper*. Yet somehow Will manages to make us aware that his role-playing is not just intended to deceive the Tibetan hill-people amongst whom he needs to pass unnoticed in order to be able to collect seeds, but is also for the benefit of the admirers of Rudyard Kipling and George Henty who demand that a plant-hunter in China should look like Alan Quartermain or Bulldog Drummond in tribal costume. At the same time, he projects a hint of irony, the amused message that whilst he's enjoying the subterfuge, we should not mistake the mask for the reality.

I kept the article, and over the following years I did my best to find out more about Will Purdom. This was easier said than done. Few of his personal papers survive, he published very little, and contemporary references to him by third parties are few and far between. More than one would-be biographer has regretfully concluded that there simply isn't enough material to enable a full account of Purdom's life to be written.

In 2015, however, I heard of a collection of Reginald Farrer papers recently gifted by the Farrer family to the Royal Botanic Garden Edinburgh which included some Purdom correspondence, and went to Edinburgh to see for myself. From Edinburgh I went to the Royal Horticultural Society archive in London to look at three surviving Purdom notebooks. I visited the archives of the Royal Botanic Gardens, Kew, and the Public Record Office, also at Kew, and searched, initially on the internet, the archive of the Arnold Arboretum in Boston. I engaged a collaborator who searched Chinese archives for me, and I procured copies of papers held in libraries in Australia. I discovered that whilst contemporary documents written by third parties and referring to Will Purdom are scarce, they do exist, and they can be tracked down and pieced together to create a whole greater than its parts.

Furthermore, the account of his life which I wanted to write was one setting Will in the context of the society and era in which he lived, both in Britain and China. I wanted to show the choices – often very limited choices – open to him during what were socially and politically tumultuous times. The latter, at least, are fully documented and I hope that the echoes of the events in which Will was caught up may fill some gaps.

I hope also to explain how Will Purdom came quite unfairly to be overshadowed in public esteem by his contemporaries Ernest Wilson, George Forrest, and Reginald Farrer. Most of all, I hope to put the record straight concerning the achievements of a man who by sheer ability and determination overcame the formidable obstacles which society set in his path and who left the world a better place for his passage.

Francois Gordon
East Malling, Kent

Acknowledgements

In researching this book, I have consulted private and public records in Britain, the United States, and (remotely) Australia and China. My debt to the archivists and librarians caring for this material is very great, and to none more so than to Lisa Pearson, Head of the Library and Archives of the Arnold Arboretum, who was more than generous in sending me digitised documents in the early stages of the project and who during my visit to Boston guided my hands to ensure I left no stone unturned within the collection of Purdom correspondence and photographs, the largest in the world, held by the Arnold Arboretum.

Leonie Paterson, Archivist, helped me find my feet in the important Farrer collection at the Royal Botanic Garden Edinburgh, which includes a large number of Purdom photographs, many of them on permanent loan from the Lakeland Horticultural Society. Lorna Mitchell, Head of Library Services, closed the circle with her quixotic suggestion in 2016 that the Royal Botanic Garden Edinburgh might be interested in publishing this book.

Liz Taylor, the Archivist of the Royal Horticultural Society, has been consistently supportive. I am particularly grateful to have been allowed to transcribe in their entirety the three Purdom notebooks which the archive holds and to have been provided with digitised copies of other documents, including the Coombe Wood ledger and the glass slides held by the archive.

Kiri Ross-Jones, Jess Conway and Kat Harrington at the Royal Botanic Gardens, Kew showed me Will's personal file from his period of employment at Kew, other files relating to Will, correspondence from the Director, the manuscript of William Dallimore's unpublished memoir and the Veitch Chelsea register.

Leyla Cabugos, the Plant Science Librarian at the University of California Davis Library, was immensely helpful in tracking down papers concerning Will's dealings with Frank Meyer. I hope that the online archive of Meyer documents which she is working to build with her colleagues will encourage future researchers to give this unfairly neglected plant-hunter the attention he merits.

Kathi Spinks, the librarian of the Mitchell Library in the State Library of New South Wales, which holds the Dr George Ernest Morrison Collection, comprising 6.5 linear metres of Morrison's diaries, correspondence and photographs, kindly checked various diary entries for references to Will. Her colleague Joanna Goh scanned correspondence, some of it irascible, all of it interesting, between Morrison, Will, and Farrer.

Will's great-nephew, Alan Purdom, insisted that I should *borrow* much of the family archive, a treasure-house of priceless information about Will. I am deeply grateful for his trust.

Several members of the Leveson family generously went to a good deal of trouble to track down family papers relating to Betty Clifton. I hope they feel that I have treated their enigmatic great-aunt fairly.

Vicky Aspin, the former keeper of the Purdom bed at Holehird, the garden of the Lakeland Horticultural Society, and currently the Lakeland Horticultural Society Archivist, has been unstinting in sharing her extensive knowledge of Will and his collections.

Sten Ridderlöf kindly provided me with photographs of *Rhododendron purdomii* taken on his 2004 plant-hunting expedition to Taibai Shan.

Arboriculturist Luke Steer spent the better part of a day showing me the trees on the Brathay Hall estate and sharing with me his insights into their history.

Wendy Hart used her genealogical expertise to help me track down several people with whom Will came into contact and about whom I initially knew only their names.

My collaborator Cheng Yi Men searched a dozen archives in China, some of which I had never heard of, for Purdom documents which he then translated for me. He photographed the Purdom Forest Park in Xinyang, and transcribed and translated what is by far the largest and heaviest document consulted in the writing of this book, the memorial stele erected in the park in 1922.

Christina Scott, Minister and Deputy Head of Mission at the British Embassy in Beijing, kindly photographed the Purdom memorial now in the Ambassador's garden.

I have visited too many other archives, often seeking just one or two documents, to list them all, but I thank all the staff who helped me search them.

I have benefitted more than I can say from the friendship and the informal mentoring and advice generously extended by Michael Charlesworth, Professor of Art History at the University of Texas at Austin and the author of the only full study of the life and work of Reginald Farrer. His contribution to any coherence or intellectual rigour which the book may display is very great: the errors are, of course, entirely mine.

My thanks to my wife, Elaine, owe nothing to form. For the last five years she has lived uncomplainingly with Will as much as with me. She has supported what have sometimes been costly trips to pursue research and has kept my feet on the ground (more or less) throughout. This book is for her.

Transcription of Chinese names

I have used the Wade-Giles system of romanisation which was current in Will's lifetime, with one exception, the city of Xi'an, home of the terracotta army, which he knew as S'ian. Where the name of a town or city has changed, I have added the new name, and in other instances I have given the pinyin transcription, save for Peking/Beijing, which is so widely known as not to need spelling out.

The place names on the two maps showing where Will was active during 1909–12 and 1914–15 are those he knew and used and the international borders and provincial boundaries shown are those pertaining at the time.

In the case of personal names, I have given the pinyin transcription, where available, after the Wade-Giles romanisation. All Chinese names are, of course, given with the family name first.

Contents

Chapter 1	Beginnings	1
Chapter 2	Apprenticeships	9
Chapter 3	Kew	17
Chapter 4	Agitation	25
Chapter 5	The Kew strike	33
Chapter 6	New directions	43
Chapter 7	Inside the Heavenly Kingdom: Paradise Lost	51
Chapter 8	Peking, and the 1909 plant-hunting season	61
Chapter 9	The 1910 season	83
Chapter 10	1911: Things fall apart	93
Chapter 11	The first expedition: a reckoning	111
Chapter 12	Loose ends in London and return to Westmorland	119
Chapter 13	Back to China	131
Chapter 14	Winter 1914/15	145
Chapter 15	The 1915 season and return to Peking	155
Chapter 16	Forestry Adviser to the Chinese Government	163
Chapter 17	Friendship or love?	185
Chapter 18	A tree falls	193
Epilogue		195
List of plates		203
Appendices:	Appendix A	213
	Appendix B	215
	Appendix C	216
	Appendix D	217
Sources		219
Bibliography		220
Endnotes		227
Index		232
Francois Gordon		252

Chapter 1
Beginnings

William Purdom, always known within his family as Will,* was born on 10 April 1880 in the gardener's cottage in the grounds of Plum-Tree Hall, Heversham, in Westmorland.† The Hall, a pleasant but not particularly large house, stood on the northern edge of the small village. His father, also William, had been the gardener there for three years. In this narrative, 'William' refers to Mr Purdom *père* and his son will be called 'Will'.

Will was the second child and the first son born to William and his wife Jane. Both his parents had been born in Grasmere, less than 20 miles north of Heversham. Typically for their generation, during their long lives neither one ever travelled further than a hundred miles from their birthplace. William and Jane were ambitious for their son, and they might have anticipated that as a teenager he would leave Westmorland for London to pursue his education.

Brathay Hall is beautifully situated between the fells and the water on the northern shore of Lake Windermere.   Author

---

* His sisters sometimes called him 'Billy' or 'Billiam' and in adult life he was 'Bill' to some of his friends.
† Now part of Cumbria.

1

CHAPTER 1

Brathay Lodge garden 1886. Will, aged six, is on the left. His mother holds two-year-old Annie on her lap. Will's brother Harry and elder sister Margaret are on the right. William stands proudly behind his family.
Purdom family papers

Brathay Lodge, the family home where Will grew up and to which he returned in 1912 after his first expedition to China.
Purdom family papers

But they could never have imagined that he would twice travel more than 5,000 miles to China to search for plants and to work there as a forester and ecologist, that his early death would be considered a sad loss to British horticulture, or that Will would be mourned by the daughter of an earl, a peeress in her own right.

In 1883, when Will was three years old, his father became head gardener at Brathay Hall in Ambleside, a handsome Georgian mansion beautifully situated at the northern end of Lake Windermere. Brathay Hall was owned by the Redmayne family, whose wealth was derived from importing silk ribbons from China for sale at outlets including their haberdashery on Bond Street, London.

The Brathay estate comprised over 300 acres, mostly wooded or laid out in a 'picturesque' landscape style, using clumps of trees and drifts of spring bulbs to enhance the magnificent views of Lake Windermere and the fells. In addition to the walled flower garden, a second two-acre walled garden was devoted to fruit and vegetables. The Hall incorporated a large conservatory in lieu of an east wing and there were separate glasshouses for vines, peaches, orchids and cucumbers as well as a greenhouse for growing tender flowering plants. William must have made an excellent impression on Hugh Redmayne to be put in charge of all this at the age of 33.

The cottage* which went with the job doubled as a gate-lodge. Typically for a head gardener's cottage, the living-room windows looked across the drive into a two-acre walled garden, so that the incumbent could keep an eye on the flower beds and the staff working there. William and Jane Purdom moved in with their children, Margaret, Will, and his younger brother, Harry. Four more children were born in Brathay Lodge over the next decade, and the Purdom family would occupy the house for 93 years.

The Redmaynes were paternalistic members of the Ambleside community, funding the building of the local church (of which, in time, William became senior churchwarden) and subsidising the local school. This did not reflect any commitment on their part to social mobility: years later, Will vividly recalled a coruscating dialogue in the early 1890s when Mrs Redmayne enquired of his father why she had not seen Will's elder sister, Margaret, for some time? On being told by her head gardener that Margaret was attending a well-known boarding school, Mrs Redmayne informed him that he "*had no business to send their daughter to such a school, that school was for the daughters of gentlemen, not for people like him and his wife*". As an adult, Will Purdom attributed his lifelong commitment to socialism to this exchange.[1]

In the 1860s a vigorous national debate concerning education had taken place in Britain. On one side were those decision-makers – apparently including the Redmaynes – who maintained that educating the children of the working class

---

* Built in 1857 for Giles Redmayne, then owner of Brathay Hall, by Alfred Waterhouse, an apprentice piece by the architect who went on to design the Natural History Museum in Kensington and who became the leading exponent of the Victorian Gothic Revival style.

CHAPTER 1

beyond elementary levels of literacy and learning to count well enough to cope with simple commercial transactions would lead to social disruption, if not revolution. They were opposed by the factory-owners and industrialists who argued forcefully that, in a world where manufacturing involved increasingly sophisticated processes, a much better-educated workforce was essential if Britain was to retain the dominance in industry and trade which it enjoyed by virtue of being the cradle of the Industrial Revolution.

Parliament eventually sided with the factory-owners and passed the 1870 and 1880 Education Acts, which established a network of School Boards, funded by local rates, to build and staff schools where there were insufficient voluntary (in practice, this meant church) schools. From 1891, free education up to the age of 14 was available across the country for all children. The effect on literacy rates, particularly in rural areas and especially for girls, was immediate and dramatic. The Purdom children were amongst the first generation to benefit from these provisions and their parents made sure that all seven of them stayed at school until they were at least 14 years old.

Will attended the Free Grammar School in Ambleside, which catered for up to 110 boys and girls. They were taught by a single teacher supported by a team of monitors, in the main older pupils, although one or two may have been students working for a teaching qualification. The compulsory curriculum consisted almost exclusively of reading, writing and arithmetic but schools had discretion to prepare individual students for examinations in one or more 'specific' subjects. Will successfully sat the botany examination, gaining an Elementary Certificate. It is unlikely that he learned much of the subject at school rather than at home from his father and from the pages of the weekly *Gardeners' Chronicle* to which William subscribed, but it is indicative of Will's determination to extract the maximum value possible from his attendance at school that he prevailed on the schoolmaster to arrange for him to sit the examination.

The issue of social mobility was at the heart of the debate about education and whether the taxpayer should fund it. Should members of the working class be permitted, even encouraged, to aspire to join the middle class of persons and perhaps, over several generations, the ranks of the gentry, or were social divisions divinely ordained and therefore immutable?

Bizarre as it may seem, this question was answered to almost universal satisfaction by a book self-published in 1859 by Scottish journalist Samuel Smiles. *Self-Help* argued that education is desirable of itself and beneficial to society and commerce. Society should enable those at the bottom of the ladder to learn, but individuals also had a responsibility to seek out the means to improve themselves and to work hard and persevere to that end. This last point was crucial: Smiles argued that anyone, without exception, could and should lift themselves out of poverty. Those who failed to do so self-evidently lacked determination and moral fibre, and were undeserving of help or sympathy.

Self-help sold 250,000 copies in Britain alone. The Mutual Improvement Movement, a loose association of societies which ran reading rooms and adult classes in the evening and sometimes on Sundays expanded to the point where it is estimated that by the 1880s up to one-quarter of male workers were members.[2] Smiles's book also spawned a whole genre of children's books in which plucky heroes and the occasional heroine drag themselves up by their bootstraps to rise above their humble origins and triumph over adversity.

The categorisation of the poor as either 'deserving' or 'undeserving' dovetailed neatly with the demand from industrialists for government support for apprenticeships and for technical training for school-leavers. Many of those industrialists were self-made men who had spotted an opportunity arising from some technical development or manufacturing process. They were happy to agree that their personal wealth reflected their own intelligence, moral integrity and strength of character, and to support limited measures to enable those amongst the poor who were, as they saw it, similarly bright and determined, to pursue their education. They saw no contradiction in simultaneously arguing that to subsidise all the poorest members of society would only lead to the totality of the latter being insufficiently motivated to work to improve themselves.

The acceptance, in one form or another, of these ideas by the majority of Britons meant that members of the working-age and working-class population were categorised as either 'stuck' on one of the lower rungs of the social ladder with no prospect of climbing higher, by implication as a direct result of their own moral turpitude, or 'improvers', persons 'striving to get on', who were seen as morally, and sometimes even eugenically, superior to their 'stuck' peers.

William Purdom, who was the son and grandson of farm labourers, saw his appointment as head gardener at Brathay Hall as putting him firmly into the 'improvers' category. His children grew up in a household where they were actively encouraged to 'better themselves', which they all did. Of his four daughters, one became the principal of a teacher training college, one the headmistress of a London school, one a secondary-school teacher and the fourth a State Registered Nurse. His youngest son emigrated and did well in Canada and another became a gold-medal-winning park superintendent in the West Country.

William himself was now in a position where he could give others 'a leg up', and in March 1891 he agreed to take on 19-year-old local boy James Abbott as an apprentice gardener. James Abbott was very old for an apprentice but he already had over two years' experience in the vicarage garden at Witherslack, less than 20 miles south of Ambleside, so his engagement at Brathay was the continuation of his training rather than the start of it.

Over the next three years, James Abbott – who had already been trained in basic tree care and topiary as well as looking after a substantial kitchen garden – rotated though the vineries and the different glasshouses, as well as caring for the very large Brathay lawns and the surrounding shrubberies.[3]

James Abbott's apprenticeship ended in May 1894, but rather than seek employment as a gardener in his own right he instead applied to the Royal Botanic Gardens at Kew for a place as a 'student gardener'.

The student gardener scheme was the creation of the second Director at Kew, Sir Joseph Hooker, who was a close friend of Samuel Smiles.[4] Student gardeners spent two years or so doing a full day's work in at least three different sections of the gardens, and in the evenings attending lectures on botany, chemistry, and practical horticulture.

Admission was by application to the Curator at Kew. Candidates had to be unmarried men aged between 20 and 25 (slightly older applicants were occasionally accepted and after 1896 so were a very few women) and with not less than five years' experience in 'good' gardens or nurseries, including experience of cultivating plants under glass, the more the better. The Curator personally vetted applications. There was a waiting list of applicants to be notified when there was a vacancy, but this emphatically did not preclude leaving a post empty if there was no current candidate meeting the standard required. Reflecting the prestige of the course, students rarely dropped out, though occasionally individuals were dismissed, usually for 'poor timekeeping'.

At the end of their time at Kew, students received a certificate of attendance,* which they could use as a springboard to apply for what today we would call 'middle management' positions. Indeed, owners of commercial nurseries, municipal gardens and the like regularly wrote to the Curator or to the Director asking for advice on filling a post about to fall vacant, when the Kew nominee was almost invariably appointed.

From the 1860s on, the Curator received an increasing number of such enquiries from colonial administrators all over the British Empire seeking to appoint directors of botanical gardens† or advisers on farming and horticulture, or in some cases requests from companies growing rubber, coffee, cotton, or tea, also seeking specialist advisers. This reflected the established Kew policy of developing the strongest possible links between Kew and botanical and horticultural establishments all over the British Empire, and especially in India.

This policy was driven by a sincere desire to maximise the benefit from useful plants to humankind, especially that portion of it living in the British Empire, but it also allowed the Kew Director to make the case to HM Treasury that work at Kew contributed directly to the financial well-being of British colonies and accordingly

---

* The debate about whether 'Kewites' who successfully complete the course should be awarded a formal degree has been ongoing for a little less than a century, during which time most Kewites have been (and remain) unenthusiastic, considering the Kew Certificate to be superior to a B.A. or B.Sc.

† To be clear, these were not ornamental parks but trial grounds, usually focussed on different types of trees, with some associated vegetable and fruit trials.

merited public funding. The same argument was also advanced by Kew to the Board of the India Office, whose funding came from the Indian, rather than the British, taxpayer, and whose Directors were concerned specifically with India and not with the Empire generally. Neither the Treasury nor the India Office were ever wholly convinced, but Kew developed strong links with plant-hunters and botanical gardens overseas and tried as systematically as possible to identify plants native to one part of the Empire which could be expected to do well in another.

The model envisaged by Kew was that specimens of potentially interesting plants would be sent from botanical gardens overseas to Kew, where they would be grown on, under glass if necessary, before being sent out to, for example, British East Africa. Exactly this happened with various species of *Eucalyptus*, which at the turn of the 20th century were brought from Australia and widely planted in Kenya and Uganda, primarily to drain swamps and thus reduce the mosquito population and the prevalence of malaria.

Forty years earlier, Kew had played a central role in a ten-year project by the British government to obtain from Peru seeds of different species of *Cinchona* trees, the source of 'Jesuit's bark', or quinine. Stripping the bark from the trees killed them, and they were becoming increasingly scarce in their natural ranges.[5] The plantations established in the 1860s by the British in India and the Dutch in Java produced quinine at a fraction of the cost of that gathered from the wild. This, coupled with a new understanding of how to reduce the risk of contracting malaria in the first place, made it possible for colonial administrators to enjoy reasonable health and a normal life-expectancy whilst working in the tropics for years or even decades, which was certainly not a realistic aspiration prior to the last quarter of the 19th century.

Another successful translocation, from Latin America to south-east Asia, followed on from the acquisition in 1876 by Kew of thousands of *Hevea** (rubber tree) seeds. As with quinine, cultivated rubber cost a fraction of the price of the wild-gathered product, and was available in much greater quantities.[6]

In 1871, against this background of close integration between Kew and the botanical gardens which had been created all over the Empire, Sir Joseph Hooker founded the Mutual Improvement Society for Kew student gardeners. Then as now, the 'Mutual' provided a forum for students to attend lectures on either 'pure science' or on practical horticulture, some of them given by Kewites working overseas who were passing through London.

In 1893 the 'Mutual' gave birth to the Kew Guild, which helped Kewites maintain contact with each other and with Kew after they had completed their studies, as well as keeping them up-to-date with horticultural activities throughout the British

---

* Today, many observers, certainly most Brazilians, would cite this as an example of 'biopiracy', the exploitation without compensation of a community's plant resources. However, 150 years ago, the Kew scientists sincerely believed that these plants were the common heritage of mankind.

CHAPTER 1

Empire. By this time, the *New York Times* could note approvingly that "*almost every British colony has a Kew man at the head of its experimental agriculture.*"[7]

Those Kewites appointed to posts overseas were never a majority of Kew graduates, but their appointment to what were high-profile and relatively well-paid jobs added to the cachet of the Kew diploma at a time when botany and horticulture had never, at least on a popular level, been more exciting. The pronouncements of successive Directors of Kew concerning the global network of botanists and horticulturalists which Kew effectively headed may have been distinctly self-satisfied in tone. But they were right to believe that they were at the cutting edge of a discipline which was having a transformative, if not entirely positive, impact on the world and on society in much the same way as was, say, engineering.[*] It is not surprising that despite the subsistence pay, there was never any shortage of applicants to train at Kew.

James Abbott's training at Brathay Hall under William Purdom met the standard required and he had excellent references from the headmaster of his former school, the vicar with whom he had begun his apprenticeship and from William Purdom: he was entered on the waiting list pending a vacancy and on 5 November 1894 he joined the staff at Kew.[8]

The conclusion of James Abbott's training left William free to take on a new apprentice. In May 1894, Will, who had turned 14 the previous month, was engaged by the Brathay estate as apprentice gardener.

Will Purdom would in due course follow James Abbott to Kew, where he would make his mark in no uncertain terms, albeit in ways which did not endear him to the Director. He would be recommended by Kew for work overseas, would personally introduce new garden plants, flowers and trees from China to Britain, and would be responsible for the translocation of useful tree species from America to China. On his death, aged 41, he would be mourned by the horticultural community as a brilliant pioneer of progress and industry who had greatly enhanced the reputation of his country in distant lands.[9]

Even an ambitious lad like Will, with a passion for horticulture and a strong belief in his own worth and abilities, could not have anticipated that he would achieve so much. But Will did recognise this apprenticeship for what it was, a classic Sam Smiles opportunity for self-improvement. There are no surviving records concerning his training at Brathay Hall, but if we judge him by results it's clear that he seized his 'chance' with both hands.

---

[*] In today's society engineers are, broadly, seen as more powerful agents of change than botanists, and this statement may strike readers as exaggerated. But without cheap rubber for tyres the bicycle would never have evolved beyond the 'bone-shaker' and the motor-car would have remained a toy for the very rich. Hydraulic machinery depends on rubber gaskets and seals. The life-saving benefits of cheap quinine were even more immediate and important.

Chapter 2
Apprenticeships

Will's apprenticeship required him to rotate through the four core areas of horticulture in which every qualified gardener was required to be competent: flowers, fruit, vegetables, and cultivating plants under glass. Apprentices were also expected to pick up the fundamentals of topiary and caring for trees. Will learned a good deal more than the basics of this last gardener's skill-set because of the nature of the Brathay Hall grounds.

The Lake District was the last part of England to lose its glacial cover at the end of the most recent Ice Age. The glaciers sculpted the hills and fells into the rolling scenery which, ten thousand years later, makes the area a premier tourist destination. But the ground from which the glaciers retreated is poor, thin, stuff liberally covered with rocky detritus of different kinds collectively known as moraine. It is not by chance that the principal form of agriculture practiced in the region is raising sheep and cattle rather than, say, growing fruit or vegetables. The large walled gardens to the north of Brathay Hall, one for vegetables and one for flowers,* were liberally anointed every autumn with horse manure. This produced a good tilth, but the ground comprising the magnificent vista between the Hall and Lake Windermere and the surrounding fells was only suitable for growing spring bulbs and trees, the latter mainly oaks (many of them today large specimens likely to be over 300 years old) and Norway maples and sycamores.†

Most of the 350-acre Brathay estate, then as now, was wooded with pine and spruce trees and a number of the relatively recently (1853) introduced 'Big Tree', then called *Wellingtonia gigantea* in honour of the Duke of Wellington, who had died in 1852. These giant redwoods, introduced to Britain by the firm of James Veitch & Sons,‡ do not grow particularly quickly or particularly tall on Cumbrian soil, but something about the local climate suits them and there are a surprising number of self-sown specimens in the Brathay estate woods. This is an unusual state of affairs in Britain and it is an intriguing thought that Will, who would introduce the 'Big Tree' to China, not as an ornamental but for timber to be used for wooden bridges and railway sleepers, first encountered it as a teenage apprentice working in the English countryside.

---
* Sadly, both these gardens have for many years been used as car parks.
† For American readers: this is *Acer pseudoplatanus*, a European maple species commonly planted in Britain and self-seeding to become naturalised in many places, especially woodland. It is unrelated to the native American plane, *Platanus occidentalis*, whose common American name is also 'sycamore'.
‡ Messrs Veitch were pipped at the post by Scottish botanist John Matthew, who sent a small quantity of *Wellingtonia* seed to his father in Perthshire four months before William Lobb, Veitch's collector, landed in Exeter with tens of thousands of seeds. But the two live saplings William also brought with him were the first *Wellingtonias* on British soil.

CHAPTER 2

More importantly for Will's arboricultural education, it is apparent that during his childhood and teenage years the woods immediately south of Brathay Hall were being planted with rare specimen trees. There are not quite enough different species of such trees for the woods to qualify as an arboretum but there is no doubt that Mr and Mrs Redmayne wanted to be able to stroll along the woodland paths and to admire a selection of non-native trees, including Noble, Norway, Douglas, and Grand firs, Western Hemlock (another large North American tree naturalised in the estate woods), three different maples, three different kinds of birch, and five different beeches. The very rare red form of cocks' comb beech *Fagus sylvatica* 'Brathay purple' is believed to have originated at Brathay as a hybrid between *F. sylvatica* 'Cristata' and copper beech, *F. sylvatica* 'Purpurea'. Judging by their size, nearly all these trees are (in 2020) between 100 and 150 years old, so that that many, if not all, of them must have been planted during William Purdom's 50-year tenure as head gardener.[*]

Will may not have been directly involved in all of these plantings, but from childhood he heard his father talk about the latest introductions to the Brathay woods, and during his apprenticeship he must have dug with his own hands the planting-holes for some of the new specimens. As a result, during the course of his apprenticeship Will learned much more than did most of his peers about trees and arboriculture, and his later career shows that he also learned to love trees above other plants.

Will also had ample opportunity to learn about cultivating plants under glass in the handsome conservatory incorporated into the main building of Brathay Hall.[†] Growing 'exotic' tender plants in heated glasshouses and conservatories was becoming increasingly popular, and this training would also stand Will in good stead.

Will's apprenticeship ended shortly after his 18th birthday, in April 1898. He wasted no time in applying to the firm of Hugh Low & Co in Bush Hill, Enfield, immediately to the north of London, and was taken on by them. The nursery had been set up 80 years earlier to bring to market plants then being introduced from Australia and South America, and was one of the largest and best-known suppliers of orchids and other exotics in Britain.

It is hard to convey to modern readers the passion with which, in the second half of the 19th century and the first decade and a half of the 20th, many wealthy Britons embraced the cultivation and collection of orchids. There were multiple reasons for this obsession which swept the ranks of the wealthy and even the merely comfortably-off like an epidemic. Of course, orchids are beautiful, but just as importantly they had to be collected from far-flung parts of Asia or South

---

[*] I am deeply grateful to the arboriculturist Luke Steer for taking me to Brathay Woods and showing me the different species growing there.
[†] The conservatory was demolished in the 1960s but the impressive stone plinth on which it was built remains.

America. They were very difficult to keep alive in Britain and even then success was often measured in months or a very few years. The owner of a conservatory full of orchids was declaring himself, and could expect to be seen as, a connoisseur of beautiful objects who also knew how to care for them.

Furthermore, whilst possession of a fine orchid collection proclaimed the owner to be a person of taste and discernment, it also affirmed the Victorian credo that Man could and should use science and engineering to control and dominate Nature, in much the same way that Great Britain controlled and dominated the Empire from which much of this natural bounty was brought to the Mother Country. Nor was it at all a coincidence that a large and diverse collection of orchids were incontrovertible proof of the wealth of their owner.

Finally, at a pinch, a fine specimen could readily be translated into cash. In 1903, for example, when the 23-year-old Reginald Farrer was invited to join his Oxford friend Aubrey Herbert on a trip to Japan which Mr Farrer *père* was reluctant to fund, he jauntily told Aubrey "*I will sell my last orchid.*"[10]

The prices which collectors were willing to pay for tropical orchids reflected this 'orchidomania'. In the 1850s, when a skilled artisan might, if fortunate in his trade and his employer, earn £3 a week, and a master craftsman in a sought-after speciality might aspire to £4 a week, a relatively common orchid could be had for £10, but to pay £50 for a single plant was not unusual. Rarities could and did sell for anywhere between £100 and £500, and once, in 1906, a unique nondescript sold at auction for 650 guineas[11] (£682.50 – the equivalent purchasing power of this sum today would be well into six figures).

At these prices, the potential profits for nurserymen able to source orchids overseas and bring a reasonable proportion of them back to Britain* alive were on a scale not seen in the horticultural industry since the Dutch 'tulip fever' of the 1630s.† The nursery owners were not slow to grasp this, and responded by recruiting and sending abroad an ever-increasing number of collectors. These 'plant-hunters' travelled to increasingly remote and dangerous parts of the world with instructions to bring back as many rare species as they possibly could, whilst at the same time sabotaging the efforts of employees of rival nurseries to do likewise. They were the first plant-collectors to search out new plants solely for commercial gain, with virtually no concern for the advancement of botanical knowledge.

The details of the struggles between rival orchid-hunters are outside the scope of this book. Mortality amongst collectors was high. Some died of fever, others were killed by hostile local people or in some cases by tigers. Some drowned in shipwrecks. Many simply set off into the jungle and were never seen again. Competition for rare specimens was so fierce that we may reasonably suppose

---

* Or France, Germany, or the USA: the demand was global, but London dominated the trade.
† In the 1880s, nurserymen learned to propagate hybrid orchids, which sold for as little as £1 or £2 each, but this had little effect on the demand for authentic species orchids, or on the prices they commanded.

CHAPTER 2

that some of the latter group ventured into territory being 'worked' by one of their rivals and were victims of foul play.*

Nothing about Will's career suggests that he was particularly interested in orchids, but Low's also had a substantial catalogue of fruit and ornamental trees and rhododendrons, a genus for which he does appear to have had an affinity. During his time with the firm, Will worked in the soft wood propagation department,[12] to which several of the four acres of greenhouses at Bush Hill were dedicated. In later life, Will was considered to be an exceptionally talented propagator, especially of woody plants, and it seems that his time at Low's marks the beginnings of his mastery of the art.

Low's nursery also deserves a mention because of the example of Smilean self-improvement set by the principal shareholder, Hugh Low, the elder son of the founder of the firm. In 1844, when he was just 20 years old, his father sent him to Borneo in search of orchids. On his return in 1847, laden with specimens, Hugh wrote a successful book about the history and culture of Sarawak.

In the same year James Brooke, the 'White Rajah' and Governor-General of the new British colony of Labuan,† offered Hugh Low the post of Colonial Secretary. Low accepted and spent the following 32 years in Borneo and Malaya, where he proved a highly successful administrator.

On the death of their father in 1863, Hugh's younger brother, Stuart Low, reluctantly abandoned his career as an officer on an East India Service tea clipper to take over the management of the company, which became known as Stuart Low and Co. Hugh supported the firm by sending orchids from Borneo and Malaya, including several previously unknown species.[13]

In his official capacity, Hugh Low, who in 1883 became Sir Hugh Low, KCMG, strongly supported research on growing, amongst other crops, coffee, tea, and rubber. He planted the first *Hevea* rubber trees to be grown on the Malay peninsula and established a model rubber plantation, giving him a good claim to have kick-started a national industry which currently produces one-fifth of the global rubber crop.

Sir Hugh maintained a regular correspondence with Sir Joseph Hooker, the Director at Kew, and on his retirement in 1899 he not only took an active interest in the family nursery but also maintained his links with Kew.[14]

Sir Hugh's papers are lost, and there is no evidence that he had any direct contact with Will, but even a junior worker at Low's must have learned a good

---

* Collectors who have murdered their competitors don't leave records of the crime for later researchers to find, and circumstantial stories take too long to recount, but in October 1878 one of Low's collectors was challenged to a gunfight by William Arnold, a collector working for Low's bitter rival Frederick Sander. They were passengers on the same boat to Caracas, in search of the same orchid, *Masdevallia tovarensis*. Low's collector, White, ran away, but the following May he despatched 80 cases of *Masdevallia* to Low. Several years later, Arnold returned to Venezuela in search of more orchids and shortly afterwards his dead body was found on the bank of the Orinoco river. Frederick Sander, who sometimes had as many as 20 collectors working for him in nine or ten different countries, saw at least eight of them die 'in the field'.
† An island off the northwest coast of Borneo, now part of Malaysia.

deal about Sir Hugh's botanical and horticultural activities, which had clearly profited both the British Empire and Sir Hugh personally. Quite how much impact this made on Will we can only guess, but he was then in his late teens and early twenties, for most of us a formative age.

After just under three years, Will left Low & Co and in the spring of 1901 went to work in the propagation department of James Veitch & Sons' Coombe Wood nursery, near Kingston-upon-Thames. This was a prestigious post: the Veitch* family were titans of the 19th and early 20th century horticultural industry and the firm was known to set a high standard.

One consequence of Veitch's stellar reputation was that when the wealthy customers who bought plants from the firm needed to engage a new head gardener, they often enquired whether there were any Veitch employees looking to move on whom the Director could recommend. This made the firm attractive to able young men who used it almost as an employment agency (the management referred to them as being "*in for a job*") and who would work there for a year or two for low wages – 12 shillings a week plus bed and board during the first year and 15 shillings thereafter (there were also some overtime payments). It also created a network of head gardeners who would almost automatically source plants from Veitch & Sons.

John Veitch had first set up as a purveyor of forest, ornamental, and fruit trees near Exeter in the late 1700s, before he and his son James bought a 25-acre site near the city centre in 1830, where James set up the Veitch Exotic Nursery. James shrewdly exploited his proximity to an important port to build up contacts with mariners and overseas residents who might be prevailed upon to send or bring him seeds and plants from America, West Africa or India, and the business prospered.

Faced with increasing competition from rival purveyors of rare foreign plants, James decided in 1840 to employ his own plant collector, William Lobb, whom he despatched to Rio de Janeiro. William spent the next three and a half years travelling extensively in South America, during which time he sent back to Exeter over 200 previously unknown plants as well as several thousand *Araucaria* (monkey-puzzle) seeds. Even before William Lobb's return, it was clear to James Veitch that despite the considerable cost of the expedition it would yield a good profit and in 1843 he sent William's brother, Thomas Lobb, to the Far East, from where Thomas sent large numbers of tropical rhododendrons, orchids, and other hothouse novelties.

Demand for the latter from Veitch's established wealthy aristocratic clients and from the burgeoning Victorian middle class continued to increase. In 1852 James Veitch bought a large plot in Chelsea, then a village just west of London, and set up the 'Royal Exotic Nursery' there, his intention being that some plant stock, including

---

* Pronounced to rhyme with 'peach'.

trees and large shrubs, would be cultivated in Exeter and most hothouse stock would be grown in Chelsea. The Chelsea nursery was managed by James' eldest son, James junior, and the Exeter one by the younger son, Robert. In 1862, on the death of their father, the brothers agreed to divide the firm's assets, Robert retaining the Exeter nursery and James Veitch junior the Chelsea business, which became the firm of James Veitch & Sons.

Harry Veitch headed the Chelsea-based nursery which dominated British horticulture for several decades before closing in 1914.

RHS Lindley Collection

In 1865, James junior signed a 50-year lease on 35 acres of land at Coombe Wood near Kingston-upon-Thames, where he grew trees and herbaceous plants. Fruit trees were grown at a smaller site in Langley and orchids in Slough.

James junior died in 1869 and was briefly succeeded by his eldest son, John Gould Veitch, who died of tuberculosis in 1870 aged just 31. Harry, the brother of James junior, then took over the running of James Veitch & Sons, which he did with conspicuous success, making a major contribution to the development of horticulture and botany and accumulating a large personal fortune.[*]

It had always been the intention of Harry Veitch – who, though happily married for many years, had no children – that one of his two nephews should take over the family business. To that end, he had devoted a good deal of anxious care to the education of his late brother's eldest son, James Herbert Veitch. In 1898 he appointed James Herbert as Secretary of the company to which he had transferred the business, making him the *de facto* Managing Director of James Veitch & Sons.

It is clear from Harry Veitch's personal correspondence that he had reservations about whether James Herbert was capable of discharging his new responsibilities to the required standard. As Harry told his old friend Charles Sprague Sargent, the Director of the Arnold Arboretum, his nephew had in earlier years, *"found routine work irksome"*,[15] and had shown a liking for spending large sums on tourism in Asia and the Far East which dismayed his uncle, who was footing the bills. Nevertheless, James Herbert was an expert botanist, with an encyclopaedic knowledge of the plants grown and sold by the family firm, and he and his younger

---

[*] As an illustration of the profits which could be made in the trade at this time, one of the first trees 'grown on' at Coombe Wood was the giant redwood, then known as *Wellingtonia gigantea* (now *Sequoiadendron giganteum*) a large quantity of seeds of which William Lobb had brought back from California in 1863. Harry Veitch claimed in a 1920 interview that the firm sold 1,000 seedlings on the first day on which they were put on the market at two guineas (£2.10) for one, or 12 guineas (£12.60) for a dozen. The seed is not hard to germinate, and if the seedlings sold about equally in dozens and individually, a reasonable estimate is that the profit realised by Veitch & Co on that one day was more than £1,000, the equivalent in 2020 purchasing power of between £350,000 and £500,000.

brother, John Gould Veitch junior, were the last surviving male members of this branch of the Veitch family. Harry swallowed his doubts, but prudently retained a majority shareholding in the company.

It soon became apparent that Harry's fears were justified. Although at first he did a good job, James Herbert began to spend more and more time working obsessively on a history of the Veitch nurseries and their plant introductions.* When he did emerge from his office his behaviour towards the senior foremen, the middle management crucial to the continued well-being of the business, caused several of them to write privately to Harry to complain that this was no way to run the firm for which many of them had worked for decades and amongst whose employees Harry had succeeded in building up great *esprit de corps*. Even worse, James Herbert's interactions with those Veitch customers who, by reason of their social status or wealth were accustomed to dealing with the head of the firm, were erratic, irrational, and all too often downright insulting, causing many of them to take their business elsewhere.[16]

Will around the time of his 21st birthday, April 1901.   Lakeland Horticultural Society

This behaviour did not reflect a deliberate policy adopted by James Herbert, but was a consequence of his poor physical and mental health. He was in the tertiary stage of syphilis, characterised by personality change and loss of social skills, irritability and hostility, confusion and forgetfulness, as well as difficulty speaking or controlling movement. All these symptoms became more severe as the disease progressed.

Although almost never publicly referred to,† this illness was far from rare in Britain at this time. Harry Veitch and many of the Veitch employees privately understood the nature of the 'nervous affliction' publicly cited as responsible for James

---

* To be fair to James Herbert, the book, *Hortus Veitchii*, was well-received on publication in 1906, and copies now command high prices.

† A rare exception came from the pen of plant-collector Ernest Wilson, who bitterly resented James Herbert's 1906 decision to have several thousand *Davidia involucrata* (dove tree) seedlings which had been grown at Coombe Wood from seed sent back from China by Wilson destroyed. James Herbert calculated that the firm would make more money by bringing only a few hundred dove trees to market and pricing them as rarities. But in his 1917 book *Aristocrats of the Garden* Wilson described this action as undertaken "*for no other reason than to satisfy the whim of an unbalanced mind.*"

CHAPTER 2

Herbert's behaviour and as justification for Harry's increasing involvement in the day-to-day running of the firm. Still, Coombe Wood cannot have been a particularly happy place when Will worked there, between March 1901 and August 1902.

In April 1901 Will went home to Ambleside to celebrate his 21st birthday with his parents and siblings. The studio portrait taken to commemorate the event shows a strikingly handsome young man, broad-shouldered, confident and elegantly dressed. His parents must have been pleased to see that working in London evidently agreed with Will, though they may have been less happy that he had become a heavy smoker, a habit he retained to the end of his life.

Despite his youth, Will's *curriculum vitae* would have impressed any prospective employer to whom he might have been recommended for a position as, for example, deputy head gardener of a large estate, with the prospect in due course of becoming head gardener, or as head gardener of a smaller establishment. If he had remained at Coombe Wood it would only have been a matter of time before he received an offer on these lines. But Will was aiming higher than that, and in June 1902 he applied for post as a student gardener at Kew.

Chapter 3
Kew

The precise terms on which Will Purdom entered into service at the Royal Botanic Gardens, Kew were to be the subject of bitter dispute over several years. It is accordingly worth looking in detail at his application in June 1902, fortunately preserved in its entirety in his personal file at Kew.

Will had kept in touch with James Abbott, his predecessor as apprentice gardener at Brathay. Abbott had done well at Kew as a student gardener, earning five certificates for his botanical studies and working his way up to the post of arboretum propagator before leaving Kew in September 1898. He was now running his own nursery business in Rushden, Northamptonshire, and was happy to provide Will with a reference for the Kew Curator. He wrote to William Watson on 14 June 1902

> *"There is a young fellow at present working in Messrs Veitch's Coombe Wood nursery who is most anxious to get into Kew. I have known him ever since he was a schoolboy and have always known him honest, steady, industrious and much interested in his work [...] I have recommended him to come over and see you personally, and I feel sure that if you give him a trial you will not regret it. His name is Wm. Purdom."*

Will did not visit Watson, but instead wrote to him on 22 June: "*As I am desirous of having a term in the Royal Gardens Kew, for the furtherance of experience, I beg to apply for the form of application …*" This letter is endorsed: "*Form sent 25.6.02.*" The use of a printed form was a recent innovation by the Director of Kew, William Thiselton-Dyer.[*] A single sheet of foolscap, it asked for name, nationality and age, height (Will gave his as 5 feet 9 inches), and the names of previous employers and the length of time with each one. Will listed the owner of Brathay Hall, Giles Redmayne, Low & Co, and James Veitch & Sons. The form was countersigned on 28 June by George Harrow, the manager at Coombe Wood.

The reverse of the form set out the conditions of service as follows:

ROYAL BOTANIC GARDENS, KEW

APPLICANTS for admission as Gardeners into the Royal Botanic Gardens are furnished with a copy of this paper, which, when filled in, must be signed by their present or last employer, and returned to the Curator, accompanied by a letter in applicant's own handwriting and with testimonials in English from employers or head gardeners. Foreigners must be able to write and speak English.

Applicants must be at least 20 and not more than 25 years of age, and have been employed not less than 5 years in good gardens or nurseries.

---

[*] The printers' marks suggest that the form was printed in April 1902.

They must be healthy, free from physical defect, and not below average height.* Whilst at work they must wear blue serge suits and grey flannel shirts with turned down collars.

The applicant will be informed if his name has been entered for admission, and, on a vacancy occurring, will receive notice to that effect. Should he not be appointed within 3 months, the application must be renewed.

The wages for gardeners are 21 shillings per week, with extra pay for Sunday duty.

Will's letter accompanying the form was brief and to the point:

"*Sir,*

Enclosed you will find the application form, filled in, giving places and period of my previous experience and also two testimonials from my present and last employers.

Being anxious to extend my experience [I] *am desirous of taking a term in the Royal Gardens, Kew.*

Yours truly, William Purdom"

Will was immediately entered onto the Kew waiting list and on 1 August 1902 the application form was endorsed by William Watson "Informed", meaning that Will had been told there was a vacancy. On 18 August 1902 Will entered into service at the Royal Botanic Gardens, Kew.

Will's admission to Kew marked a step-change in his horticultural education. At this time, the 'student gardener' training course at Kew was indisputably the best of its kind in Britain, and probably the best in the world. Then as now, there were few doors in the horticultural world which the Kew diploma did not open. 'Kewites' could realistically aspire to leave Kew for jobs with substantive managerial responsibility and with prospects for further promotion.

The Kew diploma was hard-earned, and the students' working day at Kew was long. During the week, they started work at six in the morning† (or, in winter, at dawn), and took a 45-minute break at eight. On two days a week the morning break was followed by 45 minutes 'botanising', studying plant material or herbarium specimens.‡ They then worked until noon, when they took a one-hour lunch break before working until six in the afternoon (or sunset, if earlier). After a one-hour 'tea break', on two or three evenings a week there were lectures on systematic, economic, or geographical botany (there were also optional lectures on 'British Botany') and on physics and chemistry. The students were required

---

* In 1902, the average height of British men was between five foot six and five foot seven inches.

† To be late to work by less than 15 minutes meant a reprimand, and more than three reprimands meant dismissal. Students who were more than 15 minutes late were sent home until lunchtime, lost half a day's pay, and were told that if this happened again, they would be dismissed.

‡ A herbarium is a collection of dried and pressed (or, rarely, preserved in alcohol) plant material arranged by botanical genus and species. The Kew herbarium was begun in 1852 and in 2020 comprises approximately seven million specimens.

to take notes at the lectures, which were marked by the Kew staff to determine whether or not they would receive a certificate for the subject concerned. After the lectures, they spent two or three hours studying in the library. There were weekly 'field trips' and visits to Royal Horticultural Society flower shows and four or five day-long excursions a year. On alternate Saturdays the official working day ended at one in the afternoon and there was no work on Sundays; in practice, especially in the summer, the students were often required to work on their 'free' Saturday afternoons and on Sundays, when they were paid overtime.

William Watson saw in Will both intelligence and determination to succeed. Subsequent events would amply vindicate that judgement and Will would, in every sense, go far. But, as the Kew authorities would rapidly learn, in recruiting him they had grasped a tiger by the tail.

Any modern management consultant looking at the administrative structures and staffing of the Royal Botanical Gardens at Kew in the early years of the last century and the global role and responsibilities discharged by Kew – in the main, to a high standard – would immediately conclude that it's impossible that it could ever have worked. The professional staff at Kew, excluding labourers, carters, cleaners and the four-man Kew Constabulary, but including 50 or so student gardeners, never reached three figures. Prior to 1914, the annual budget hovered a little over £30,000, with the occasional supplementary grant for a capital project such as a new building. With these resources, the Director was required to manage a 300-acre estate and public park; to maintain and continue to develop the (then) largest herbarium of botanical specimens in the world, making it available to scholars and facilitating publication of their work on classification and taxonomy; to educate the wider public about the potential benefits to be derived from botanical discoveries; to train young botanists and horticulturalists, and to support and, as necessary, direct botanical work across an Empire which at this time covered only slightly less than a quarter of the global land mass. How on earth was it done?

A large part of the answer lies in the intellectual climate and the cultural values arising from the Scottish Enlightenment and the Industrial Revolution. Educated persons and decision-makers in government did not see the tasks listed above as a set of discrete functions, but as the enumeration of different aspects of a mission which for nearly all of them was a sacred obligation, namely to use the power of rational thought and the possibilities created by modern technology to benefit humanity, and especially the one-fifth of it living in the British Empire. The consensus to this effect united those who saw this task as their Christian duty and agnostics who embraced the Utilitarian philosophy articulated by Jeremy Bentham and John Stuart Mill.

One consequence of the above for the Director of Kew and his senior staff was that their positions vested in them considerable moral authority, which they were able to translate into action on the ground to an extent which could rarely be achieved today.

Thus, the Kew Director *expected* to be consulted about the appointment of senior staff to botanical gardens and proving grounds across the Empire, and absent quite extraordinary circumstances he *was* consulted. Further, when he recommended a person for a particular post they were invariably appointed. The complete absence of a formal nexus between Kew and, say, the body responsible for running and funding the botanical gardens in Jamaica, was seen by all concerned as wholly irrelevant, and the absence of any overarching administrative structure encompassing the network of thirty-odd botanical establishments around the Empire did not diminish its effectiveness. Indeed, it may to a degree have enhanced it. For example, when Kew asked the Calcutta Botanic Gardens to send seeds or living plants, or *vice versa*, there was no bureaucratic requirement for approval or clearance – unless, of course, the request had substantial financial implications, in which case Kew would probably have obtained the necessary approval in advance.

All this meant that when Will joined the Kew staff he was entering a community which took great pride in the leading role it played in practical 'economic botany', translocating plants around the world to where they believed it would do the most good.

As Will rotated through the different departments at Kew he found himself handling plant material quite beyond his previous experience, which had largely focussed on decorative plants. Although the Kew propagation beds and glasshouses did contain previously unknown ornamentals sent by plant-hunters from China or Latin America to the nurseries who sponsored their expeditions and which were passed to Kew in exchange for their identification or classification, they also contained plants being assessed or grown for their utility in different parts of the Empire. Examples of the latter included the very fast-growing *Eucalyptus grandis*, propagated by Kew in hundreds and thousands for despatch to Africa and India, where it both drained swamps and provided a good source of timber for construction or export, and camel-thorn, *Alhagi maurorum*, procured by Kew from Cairo and destined for Australia, where the authorities needed fodder for the camels increasingly used there as beasts of burden.*

The 1903 Kew course of lectures on 'Geographical Botany' covered the practical benefits of plant translocation and current 'best practice' for forestry. One aspect of this was the alleged connection between forest cover and rainfall, first proposed over two thousand years ago by the Greek philosopher Theophrastus, 'the father of botany'. By the 18th and 19th centuries a substantial majority of scientists believed that forests create rain within a locality, and that the converse is also true, *i.e.* that large-scale deforestation will reduce rainfall in the immediate region, causing the soil to dry out, perhaps very severely. Hence, in Britain and the US these ideas were labelled 'the dessicationist theory'.

---

* Alas, camel-thorn is now legally proscribed in Western Australia, listed amongst the most noxious prohibited weeds.

At its simplest, it was thought that forests increase rainfall when tree roots draw groundwater from the soil, which the leaves then release into the air as water vapour. Given enough trees, the vapour forms clouds which then shed the water, downwind, as rain. The assertion by A H Unwin in his 1920 work on West African Forestry that *"according to the figures of Indian rainfall statistics [...] forest cover increases rainfall by 16 to 28 per cent"*[17] is representative of thinking in the late 19th and early 20th century.

President Theodore 'Teddy' Roosevelt, a dedicated outdoorsman, was a firm believer in the theory, which was a major driver for his vigorous support for the establishment of the US Forest Service and for shifting 63 million acres of public land into the national forest system. In 1908, Roosevelt met Frank Meyer, who had recently returned from collecting in China for three years on behalf of the US Department of Agriculture, for a briefing on the damage caused by deforestation in China. Roosevelt was so impressed with the briefing that he took the unprecedented step of attaching a number of Meyer's photographs to his eighth annual 'State of the Union' message to Congress to illustrate the dire consequences which would flow from a failure to preserve the forests of America.[18]

Belief in the dessicationist theory greatly reduced in the course of the 20th century, after large-scale planting of trees in South Africa, Australia and French West Africa failed to produce the hoped-for increase in regional rainfall. But it is quite clear from his later writings that Will firmly embraced the theory, and, like the great majority of trained foresters of his time, was convinced that planting very large quantities of trees, especially in areas which had in the past been covered in forest, would, in time, change the local climate for the better.*

The Kew authorities recognised that Will was a highly skilled propagator. On 6 March 1904, he and William Dallimore, the foreman† of the arboretum propagation pits, where trees and woody plants were propagated, jointly gave a demonstration of propagation to the student gardeners of the Mutual Improvement Society. In April, 19 months after arriving at Kew, Will was appointed sub-foreman at the Arboretum, the top post open to a student gardener, and at the end of the year he won the Sir Joseph Hooker prize for the best lecture given to the student gardeners.

William Dallimore, nine years Will's senior, and himself a former student gardener, would become a leading authority on trees and silviculture in Britain, the author of a best-selling book on British trees and, much later, the creator of the Bedgebury Pinetum. Will could not have hoped to find a better mentor to help him develop his knowledge of trees and arboriculture.

---

* After 60 or 70 years during which this theory was roundly rejected by climatologists, the current view is that in certain limited circumstances, including particular combinations of forests and high mountain ranges which effectively trap rain clouds, forests may indeed enhance precipitation. But the key point is that Will, entirely reasonably given the then state of scientific knowledge, believed the theory to be sound and universally applicable.

† A job-title which did not reflect the managerial responsibilities of the position, the bottom rung of senior management at Kew, and which a few years later was changed to 'Curator'.

CHAPTER 3

If the staff at Kew were apostles of scientific and economic botany, those concerned with trees and forestry were the evangelists. The huge increase in the global demand for timber triggered by the Industrial Revolution and the proliferation of railways had led to the felling of vast areas of forest. This had in turn led to growing concern about the future availability of a strategic resource, and about the impact which clearing the forest was having on local groundwater, including floods caused by increased run-off, and on climate. The scientists studying these phenomena were advancing arguments which embraced what today we would call anthropocentric changes to natural cycles, making them by any reasonable definition early ecologists and climatologists. Their conclusions were fiercely disputed by those whose short-term interests would suffer if they were, for instance, prevented from clear-felling their land or forced to re-plant after clearing the forests.

William Thisleton-Dyer, Director of Kew 1885–1905, in the uniform he designed for himself as the 'Inspector of the Kew Gardens Constabulary'. He wore it to work at least one day a week.

Royal Botanic Gardens, Kew

In short, Will's study of forestry and silviculture at Kew concerned not only the taxonomy of trees and the theory and practice of tree care, but was at least equally about the highly topical political questions arising from the ongoing and vigorous public debate about deforestation and reforestation. At the time the tide of debate was running strongly in favour of those who argued that the wider benefits to society which flowed from strictly enforced controls on felling and an interventionist policy of reforestation trumped the rights of land-owners to do as they pleased with their property. Certainly, this was the position taken by William Dallimore. Will's later career permits of no doubt that he was firmly of the same mind and, as we shall see, his Kew training in arboriculture and forestry changed the course of his life.

However satisfying Will found his work and training at Kew, he suffered along with the rest of the staff under the petty tyranny of the Director, Sir William Thiselton-Dyer, the son-in-law of the second Director of Kew, Sir Joseph Hooker, whom he had succeeded in 1885.

In justice to Thiselton-Dyer, it has to be said that he was well-qualified for the post, having served as Assistant Director for ten years. His name is esteemed amongst botanists concerned with plants from the Indian sub-continent because of his work in editing the later volumes of *The Flora of British India* and amongst all taxonomists as the first editor of the *Index Kewensis*, initiated by Charles Darwin to register the botanical names, by species and genera, of all seed plants.

Thiselton-Dyer was a successful botanist, then, but the title of the chapter in the official history of Kew concerning his directorship, 'An autocrat at Kew', barely does justice to his managerial style.[19] He could not brook being in any way contradicted, and was *"obsessed with the idea that he could only govern by exacting abject submission from those beneath him, which sometimes took the form of humiliating his principal officers before their subordinates"*.[20]

Furthermore, his preoccupation with uniforms fell only just short of mania. On appointment as Director, he *ipso facto* became head of the four-man Kew Gardens Constabulary, whereupon he decided that this entitled him to call himself the Kew Inspector of Constables and to wear a uniform he had designed himself and which he wore to work at least one day a week.[21] He forced all the Kew officials to join the Kew Gardens Fire Brigade and attempted to insist on their wearing uniform for weekly drills, but they rebelled.[22] He then spent 24 years trying to introduce uniforms for all the Kew staff but the Board of Agriculture, the British government department responsible for the funding and strategic direction of Kew, vetoed the idea on grounds of cost. Thiselton-Dyer had instead to content himself with a dress code so strict that he could, and did, dismiss a gardener for wearing a bowler hat to work – bowler hats were the sole prerogative of foremen.

The kindest description possible is that Thiselton-Dyer personified Victorian – *early* Victorian – social attitudes at precisely the time when the accession to the throne of Victoria's son 'Teddy' was seen by many of his subjects as a welcome opportunity to move on from the excessive formality and pitilessly rigid social hierarchy of his late mother's reign. The stage was set for conflict at Kew, and in 1905 it erupted.

Chapter 4
Agitation

The labourers and student gardeners at Kew had for some time been unhappy with their pay, in 1904 24/- (£1.20) a week for the former and 21/- (£1.05) for the student gardeners, with a small supplement when they worked on their 'free' Saturday afternoons or Sundays.

The wages paid by Kew were close to the average rate for gardeners in Britain: the British Gardener's Association, created in June 1904 partly at the initiative of William Watson, Curator of the Kew Gardens, had surveyed 75 gardens all over the country and established that the average wage paid to a man who had trained for five years was 17/- (£0.85) a week with a bothy (accommodation) or £1 without. Employers were easily able to fill vacancies at these rates because of the very high levels of unemployment amongst qualified men. This may explain the modest goals set out in the British Gardener's Association prospectus dated 1st June 1904, namely to achieve the adoption of a scale of wages set at 21/- a week without accommodation for journeymen gardeners and 24/- for a foreman/head gardener if provided with a house, or 27/- a week if not.[23]

Because of the huge changes in the relative price of goods and services over the last century it is almost impossible to translate these sums into modern money, but in 1904 a serge suit cost at least £1 and a stout pair of boots 5/- (25p). A room in a poor-quality lodging-house in or near Kew cost 7/6d (37.5p) a week.[24] Student gardeners hired themselves out during their limited spare time to look after private gardens in Kew and sometimes shared a room to save money. In 1907, for example, Will shared a room with William Johns, an old friend with whom he had worked at Low & Co and who came to Kew as a student gardener in February 1907.

Despite such shifts, money was tight for the Kew student gardeners, and the labourers, some of whom were married with children, must have found it almost impossible to make ends meet. A recurring theme in descriptions of Will's character by those who knew him is his unfailing sympathy for anyone suffering unfair or unjust treatment and his refusal to tolerate such behaviour in others. When William Watson looked for support for an approach to the Director under the auspices of the British Gardeners' Association to seek improvements in the conditions of service of the Kew workforce, he didn't have to ask Will twice.

Another grievance of the student gardeners was that the syllabus of the course of lectures they attended in the evening did not adequately reflect developments in horticulture and botany over the last 30 years, potentially putting them at a disadvantage when competing for jobs with graduates of technical colleges with more up-to-date courses. Several petitions asking for reform had been presented by the student gardeners to the Director but had been rejected.

The British Gardener's Association represented an attempt to create a professional body which would establish a register of gardeners *"with a view to regulating and controlling the labor* [sic] *market for gardeners"*. In a deliberate attempt to avoid the Association being identified as a trade union, its constitution mentioned the promotion of the interests of the profession, but not as the Association's principal objective. Thiselton-Dyer was unimpressed with this sophistry and when, in early 1905, the Association organised the presentation of petitions from the foremen, student gardeners and labourers, instead of forwarding them to the Board of Agriculture and Fisheries, he summoned them to separate meetings.

The demands the garden staff set out in their petitions were scarcely incendiary. The labourers wanted an extra 4/- (20p) a week pay, and presented detailed figures to back up their claim that the cost of living in the vicinity of Kew meant that 25/- a week was the minimum they needed to live decently. The student gardeners wanted the same pay as the labourers, and improvements in the course of lectures. The foremen wanted an increase on the same lines as the student gardeners and labourers, to change their status of 'unestablished' civil servants to 'established', so that they would receive a pension on retirement, and a change in their job-title to 'Assistant Curator', which would more accurately reflect their responsibilities and would be helpful to those applying for similar jobs outside Kew.

Thiselton-Dyer was having none of it. He told the labourers that if they were unhappy at Kew they could leave – he would, he told them, easily fill their places if they did so – and the foremen and student gardeners that he was shocked at their lack of gratitude for their training at Kew. He refused to sponsor their request for a change of job-title, a change in their status within the Civil Service, or increased wages. William Dallimore records that after the meeting the labourers were in a very angry mood, and that he and the other foremen *"left the office feeling like reprimanded schoolboys."* [25]

Dallimore himself considered leaving Kew and setting up on his own account as a nurseryman, and Watson pressed him to accept an offer to go into partnership with the son of a friend. Dallimore declined, which annoyed the Curator because *"he wanted to appoint a certain sub-foreman to a foreman's post, and mine [i.e. Dallimore's] would have been the most suitable for the man."* [26]

Dallimore was foreman at the Arboretum, which in 1905 had only one sub-foreman, Will Purdom. Watson was hoping to have Will promoted to a post which was not time-limited, as were those of the student gardeners and sub-foremen. It seems likely that Watson wanted to secure a permanent post for Will in order that the two men could continue to work together to promote the interests of the

Kew workers without having to worry about the precarity of Will's employment status. If Dallimore had obligingly vacated his position before mid-February 1905, Watson's recommendation of promotion for Will would almost certainly have been accepted, but by the end of February and for the rest of his time at Kew any such proposal would undoubtedly have been vetoed by the Director.

After their bruising encounters with Thiselton-Dyer, the Kew workers needed little persuasion to set up their own trade union. On 10 February, nearly all the staff at Kew met at the 'Kew Boathouse', a tavern just outside one of the garden gates, and resolved to establish a subsidiary of the Government Workers Federation (GWF) trade union, to be called the Kew Employees Union.* Will and a Kew labourer called William Paine were mandated by their workmates to take forward the formation of the new union. It is not clear why the choice fell on them: there were many men present who were older than Will and who had worked at Kew for longer than he had. Perhaps they were impressed with Will's grasp of the practicalities of setting up a trade union, but the vote clearly reflected faith in his and Paine's ability and integrity.

Will and Paine immediately made contact with GWF officials and attended the tenth annual conference of the trade union before visiting the House of Commons to meet Labour Members of Parliament, who encouraged them to go ahead with establishing the Kew Employees Union. The MPs even arranged a brief meeting with Lord Onslow, the President of the Board of Agriculture and Fisheries, who promised Will and Paine that he would help them improve conditions for Kew workers. The self-confidence displayed by both men in meeting a senior member of the Cabinet to make their case is remarkable, especially when one remembers that Will was not yet 24.

On 24 February, a second meeting of the Kew Employees Union adopted the constitution of the union and voted to elect Will Purdom as Secretary and William Paine as President. Will gave the meeting an account of his and Paine's meetings in Parliament, including *"the promise of Lord Onslow"* that he would support the union.

In early April 1905, Will, as Secretary of the Kew Employees Union, sent Lord Onslow a memorandum concerning the pay and conditions of employment of gardeners, labourers and constables at Kew. This was returned to Will on 6 April by the Secretary of the Board of Agriculture with a request that it be submitted *"in the prescribed manner, through the Director of Kew"*. Will duly passed the memorandum to William Watson, the Kew Curator *"to be submitted* [via the Director] *to the President of the Board of Agriculture"*, and on 14 April Watson forwarded it to Thiselton-Dyer, who immediately summoned Will.

According to the note Thiselton-Dyer penned to the Board of Agriculture immediately after the meeting, he began by informing Will that he was refusing to

---

* According to the *Pall Mall Gazette* of 6 November 1903, every single Kew employee below the rank of foreman joined the union, a telling indication of the dissatisfaction felt by the Kew staff.

receive the memorandum because *"I am not prepared to recognise any body which comes between me and my staff or professes to speak on their behalf"*. Sir William added that the stokers, constables and museum porters were clearly satisfied with their conditions of service because they had not complained directly to him, and that in any case the labourers, constables, and gardeners could not be treated on a common basis. He pointed out that the labourers had petitioned for an increase in pay earlier in the year and that this petition was under consideration at the Board of Agriculture, and told Will that there was no conceivable justification for Kew student gardeners to receive the same wages as gardeners employed in London public parks by the London County Council.

Sir William then asked Will to justify the *"trade union rate"* (30/- a week) demanded by the Kew Union. Will replied that this was the rate determined by the British Government Workers Federation, an answer Sir William described as *"absurd"*, before pointing out that the rates of pay at Kew were exactly the *"trade union rate"* set out in the prospectus of the British Gardener's Association.

In conclusion, Sir William returned to Will what must by now have been the rather dog-eared memorandum, adding that if it were re-written *"in a proper form"* (i.e., addressed to the Kew Director, rather than to the President of the Board of Trade), then it would be received. Finally, Sir William informed Will that he regarded any direct communication with the Board of Agriculture as a breach of discipline, and that he would be posting a notice to the staff to that effect.[*]

In reporting the above to the Board of Agriculture, Thiselton-Dyer could not resist a swipe at William Watson, whom he considered partly to blame for *"the current ferment"*, and a barely veiled criticism of the government of the day, which he suggested was adopting, through the Office of Works, a *"socialistic"* scale of wages, as compared to the *"economic"*[†] wages paid at Kew.

The conclusion to Thiselton-Dyer's note, asserting that *"the government must choose whether to run with the hare or the hounds"* is unlikely to have endeared him to Lord Onslow, but he ordered his civil servants at the Board of Agriculture to make a detailed survey of the wages paid in public parks, large private gardens and the horticultural industry to determine whether the wages at Kew were on a par with those paid elsewhere.[27] This took several months, and the Board then submitted the results of the survey to HM Treasury, with a request for Treasury approval for increases in pay for some Kew staff.

On 29 May 1905, William Watson used the occasion of the annual dinner of the Kew Guild, of which he was President, to urge that the Guild should expand its remit both to promote the welfare of its own members and that of professional

---

[*] Which he did, the same day.
[†] At this time, an 'economist' was a frugal individual skilled at making money go further, not a student of supply and demand and associated social and political phenomena. Thiselton-Dyer was boasting that Kew paid the minimum wage consistent with securing the necessary labour.

gardeners generally, and might even go so far as to *"show a sympathetic interest in any body meant to benefit horticulture"* – a transparent reference to the British Gardener's Association.

This provoked a furious riposte from Thiselton-Dyer, who argued in a letter to the *Gardeners' Chronicle* of 10 June that the Kew Guild (of which, he claimed, not entirely accurately, to have been the progenitor) was an institution of a purely private character, *"almost a family affair"*, which should at all costs avoid *"dabbling in horticultural politics"*. He justified this on the basis that Kew was a government establishment, which precluded its staff from taking action which might embarrass the government and concluded with the rousing assertion that *"When the gardener is worth more, he will get more"*.

In the 17 June correspondence column of the *Gardeners' Chronicle*, a correspondent writing under the name of 'Old Kewite' (there is no evidence as to his identity, but William Watson is the prime suspect) took up the cudgels, pointing out that the great majority of Kewites did not work for the government and concluding *"I gather ... that Sir Wm. Thiselton-Dyer considers that the remuneration of gardeners is [...] commensurate with their ability. That being so, the latter must be about equal to the skill of an ordinary labourer"*.

On 21 June, just after the publication of the letter from 'Old Kewite', the Treasury confirmed that, although the petition from the labourers for 24/- a week (an increase of 3/- a week) was rejected, their pay could be raised to 23/- a week. There were similar increases for the carters, horse-boys, stokers, and the constables, but *"as regards the Sub-Foremen and the Journeymen Gardeners, no increase in pay is recommended"*.

No one examining the meticulous analysis of wages in gardening and horticulture carried out by the Board of Agriculture can dispute the conclusion of HM Treasury that the rates of pay at Kew reflected the reality of gardeners' wages elsewhere in Britain. Since the repeal of the Corn Laws in 1846 and, more importantly, the huge reduction in the cost of shipping grain from America to Britain due to vastly improved merchant ships, jobs in British farming and horticulture had reduced by at least a quarter. In 1873 it had cost £3.35 to ship a ton of wheat from Chicago to Liverpool, in 1884 £1.20. Over the same period, grain prices in Britain almost halved, and by 1885 over a million British acres which had been under the plough in 1870 had reverted to pasture.[28].

The effect on rural communities was brutal: *"for twenty years the only chance for any young or enterprising person in the countryside was to get out of it"*.[29] Nearly one million British country-dwellers emigrated, and over 100,000 men and their dependants moved to the cities in search of work. Since agricultural work was what they knew, many of them gravitated towards gardening. The consequent overcrowding in the profession – two separate letters in the *Gardeners' Chronicle* on 17 June 1905 asserted that every advertisement for a gardener would attract

CHAPTER 4

well over 100 applicants – meant that wages across the country were barely at subsistence levels. This was, of course, precisely the issue Will and the Kew Employees Union sought to address, but as far as HM Treasury were concerned, 21/- a week was 'the going rate'.

On the face of it, then, by mid-1905 the Director had decisively won the battle with the Kew Employees Union, who were reduced to briefing sympathetic journalists about the low wages at Kew compared to those paid by the London County Council and – an argument which gained slightly more traction in the press – the unfairness inherent in gardeners, skilled men, being paid less than unskilled labourers.

This was a dispiriting period for Will, but he held firm to his belief that workers needed to work together to improve their pay and conditions. On 17 July 1905 he wrote to the secretariat of the Labour Representation Committee asking for a copy of the report of the last LRC Conference and for particulars regarding affiliation to the Labour Party so that this could be considered at the next meeting of the Kew Employees Union. Will also sent off a postal order for four tickets to a Labour Party reception at the London Horticultural Hall on the 19th. The Secretary of the LRC replied by return of post expressing pleasure that the Kew Union was considering affiliation and told Will that *"I have no doubt whatever but that you will find the work of the Labour Party in Parliament of great assistance to you"*.[30] Unfortunately, this exchange of letters is all that is preserved in the Kew Employee's Union file in the Labour Party archives, but it is apparent from subsequent events that the Union did affiliate to the Labour Party and before the year was out Labour MPs in Parliament would indeed be of considerable assistance to the Kew Union and to Will personally.

We know very little about how Will spent his limited leisure time whilst he was working at Kew, but in the summer of 1905 he went to the Naval, Shipping and Fisheries exhibition at Earls Court, a miniature World Fair organised to celebrate the centenary of the Battle of Trafalgar. Visitors could see the much-admired diorama of the Battle as viewed from the poop of HMS *Victory* or view the spectacle *"With the Fleet"* representing life on a Royal Navy cruiser travelling from Britain to the Mediterranean. A group of authentic Peterhead fisher-folk were shown mending nets and knitted sea stockings in a diorama of a Scottish fishing village. Visitors could also see a *"Red Indian village from Forest Lake, Canada"*. Will was accompanied by Jack Cotter, a labourer attached to the Arboretum, who was so enchanted by the spectacle that he returned several times before the exhibition closed at the end of October.[31]

The Kew Employees Union continued to lobby sympathetic politicians, notably William 'Will' Crooks, the former Mayor of Poplar and prominent campaigner against poverty and inequality, who was one of just five Labour MPs in the House of Commons. On the evening of Friday 27 October, Crooks and another Labour MP spoke at a public meeting of the Kew Employees Union, pledging Labour Party support for efforts by the union to persuade the Board of Agriculture to improve wages at Kew. This was more than Thiselton-Dyer, who may have been amongst the crowd on Kew Green listening to the speakers,[32] could bear. The following day George Merriman, the new President of the Union, and Will Purdom, the secretary, were dismissed for 'agitating', effective from 4 November.

No papers relating to what happened over the next ten days have been preserved in either the Board of Agriculture and Fisheries files at the Public Record Office or in the archives of the Royal Botanic Gardens. But we know that Will Crooks was outraged by the Director's action, all the more so since he himself had, 30 years previously, been sacked for his political activities. Crooks met the Secretary of the Board of Agriculture to demand that the decision be reversed. He was supported by the Government Workers Federation, who pointed out that the workers at Kew enjoyed the same rights as other civil servants, including the right to join a trade union and to do as they pleased with their time outside working hours. The Board could not demur, and the Kew Director was instructed to reinstate Purdom and Merriman immediately, without loss of pay. Faced with this humiliating public reversal, Thiselton-Dyer resigned. Will Purdom and George Merriman were reinstated on 13 November.

Chapter 5
The Kew strike

On 5 December 1905, the Conservative Prime Minister, Arthur Balfour, resigned, and Edward VII invited Liberal leader Henry Campbell-Bannerman to form a minority government.

Lord Onslow, a Conservative peer, resigned as President of the Board of Agriculture and Fisheries, but not before he had approved the appointment of Colonel David Prain, the Superintendent of the Royal Botanic Gardens at Calcutta as the new Director at Kew. Lord Onslow was replaced by the Liberal politician Lord Charles Carrington.

Thiselton-Dyer remained in post during December to allow his successor time to make the journey from India. Prain formally assumed the Directorship on 2 January 1906, although Thiselton-Dyer acted as his unofficial adviser for a further three months whilst Prain familiarised himself with his new responsibilities.

Prior to his appointment to Kew, David Prain had spent 20 years at the Calcutta Botanical Gardens, first as Curator of the herbarium, and since 1898 as Director. He held the rank of Lieutenant-Colonel in the Medical Service of the Indian Army, and even after his appointment to Kew signed himself 'Col. Prain'. Thiselton-Dyer had worked closely with Prain on *The Flora of British India* and had a high opinion of his successor's botanical knowledge.

Prain was pleased to be appointed Director at Kew, but he also understood that his first challenge was going to be to rebuild morale there, and that this would not be easy. Will's thoughts are not recorded, but he was surely happy to see the back of Thiselton-Dyer.

The minority Liberal government rapidly called a General Election over the period 12 January to 8 February 1906. The Liberals won decisively, securing an overall majority of 125 in the House of Commons. The

David Prain, Kew Director 1905–1922, who was determined that Will, and all the other 'student gardeners', should leave Kew at the end of their training. National Portrait Gallery

CHAPTER 5

Conservatives won just 156 seats, their worst ever result. Labour, led by Keir Hardie, won 29 seats.

Despite his privileged background, the new President of the Board of Agriculture, Lord Carrington, was known to favour a programme of draft legislation aimed at relieving the plight of the poorest in society, which was, broadly, the issue on which the Liberals fought and won the 1906 election.* On his appointment, the Secretary of the Government Workers Federation, James King, immediately approached Carrington and asked him to receive representations from the Kew gardeners, and Carrington agreed.

On 13 January 1906, therefore, Will sent Lord Carrington a neatly typed four-page memorandum, on behalf of the Royal Gardens Kew Employees Union:

"to lay before you the following facts which plainly show how inferior are the conditions of employment at Kew compared with those prevalent under the London County Council, and to inform you that, generally, the conditions of employment at Kew are exceptionally unfavourable compared to those prevailing in the Public Parks and Gardens in the London District."

The following two-and-a-half pages – densely packed with columns of figures comparing and contrasting the pay, hours of work, overtime, leave, sick pay, and superannuation of the whole range of Kew employees with those paid to their LCC homologues – are not readily summarised, but on their face support the proposition that the men and women at Kew were underpaid. The memorandum concluded with a flat statement that *"the majority of the Kew employees do not receive a living wage"* and ended with the text of a resolution passed at the last Kew Union meeting urging the Board to introduce a minimum wage of 30/- (£1.50) a week, and a 48-hour week.†

The Board of Agriculture promptly passed this hot potato to the new Director at Kew for his comments. Prain – who had been in post for a fortnight, and who might have been forgiven for muttering to himself the words 'poisoned chalice' – took six weeks to respond. When he did, it was with a twenty-page manuscript *"to show synoptically what the Kew labour force ask for, what they receive, and what it seems advisable that they should receive"*.

Prain's February 1906 memorandum said as much about him as it did about the labour force at Kew. He began by nailing his colours to the mast: as he had previously opined, the request of the *"young journeymen gardeners"* for 24/- a week was not justified. The Kew men, unlike the LCC Parks journeymen gardeners, were *"advanced students of horticulture, serving at Kew for a definite and limited period as improvers in their particular craft"*. The 21/- paid weekly to the young gardeners at Kew was *"not a wage but a subsistence allowance"*.

---

* The principal slogan of the Liberal campaign had been *"Big loaf, small loaf?"*, on the basis that if the Conservatives won they would introduce tariffs on wheat and other foodstuffs, when the price of food would rise and although the price of a loaf of bread would remain constant, the loaf would become smaller. Working-class voters found this argument compelling, hence the Liberal landslide.

† The Kew student gardeners' basic working week, excluding breaks, lectures and study time in the library, was 56 hours.

He concluded by observing that he had

> "supervised [in India], over twenty years, more Kew-trained gardeners than any other officer. I have thus had especial opportunities [...] of deciding what particular type of young gardener is most useful. The man whom a colony wishes to avoid is the man whose ambition is limited to the receipt of his pay [...] The man who is desired is the man who is ready to subordinate his personal comfort to the public efficiency, who is prepared reculer pour mieux sauter,* who does not despise the day of small things."

Low wages discouraged *"those who would be satisfied with a colourless career"*, but not *"those of a higher and more reflective type of which the Empire stands in need"*.

In short, Prain argued that the current scale of payment for young gardeners attracted morally upright persons whose willingness to embrace poverty during their time at Kew in order to improve their chances of future service in the Empire overseas proved that they were what we might today call 'the right stuff', and that an extra 3/- a week would be morally corrupting for the individuals concerned and, ultimately, bad for Britain and the Empire.

The extent to which Lord Carrington and the new Secretary of the Board of Agriculture and Fisheries, Sir Thomas Elliott, were convinced by these arguments is doubtful, but they agreed Prain's detailed proposals for the Kew staff budget, which amounted to implementing the wage increases already endorsed by HM Treasury, principally benefitting the garden labourers, plus a small increase in the number of stokers for the Palm House, and a scheme to furnish gardeners with regulation flannel shirts and serge suits on joining the Kew workforce.

In June or early July 1906, Sir Thomas Elliott passed a copy of Prain's memorandum to Will Crooks, who in turn passed it to the Kew gardeners for their comments, which on 12 July Crooks conveyed to the Board. The comments took the form of two memoranda, from which Crooks prudently cut the signatures before sending them to Sir Thomas.

The first memorandum, three closely typed pages criticising the training at Kew, arguing passionately that the young gardeners at Kew should be paid a fair wage, and lambasting the Director for apparently believing that *"the half-starved man is better than the fed"*, concluded that in considering the Director's report it should be remembered that *"he is little more than a stranger to Kew"*. The second, hand-written, began *"I have read the memorandum of Col. Prain and think it is sheer humbug"*, before citing detailed figures about the posts to which student gardeners moved on after Kew to refute Prain's arguments that the Kew diploma was a passport to a well-paid job in an Imperial possession.

Crooks' action in removing the signatures makes it impossible to be certain of the authorship of the memoranda (which was precisely his intention) but it is

---

* This French idiom is not easy to translate, but the sense of it is that taking a step back before tackling an obstacle is often a shrewd move.

likely that the first note was typed by Will and the second was written by Watson, who was one of the few Kew staff members with access to the detailed figures cited. In any event, Prain rapidly became aware that his memorandum had been passed to the Kew employees and had gone down very badly. He complained to Sir Thomas Elliott about what he considered to be a breach of confidence. Sir Thomas apologised, saying that he felt badly let down by Crooks and that this would not happen again.

On the ground at Kew, the Director's weekly tours of inspection of the grounds, invariably followed by a shower of irritating notes to staff requiring them to rectify minor instances of untidiness or clutter, ceased. This was partly out of consideration for the garden staff and partly because, as his subsequent career would show, Prain was not personally very interested in the gardens. His focus was firmly on the scientific and botanical work of Kew, and on running the Imperial horticultural training establishment. In any event, the garden staff welcomed the change of regime. Most of the workforce had won at least part of what they had requested in the Kew Union memorandum, leaving only the student gardeners' grievances unaddressed. Despite the ill-feeling generated by the circulation of Prain's memorandum, overall tension between staff and management reduced.

By the autumn of 1906, then, Prain could reasonably feel that he had a firm grip on his new post, and on the administration at Kew. He now decided that the rule that student gardeners should stay at Kew for two years should be more strictly and more consistently applied than it had been in the past, starting with the discharge of those student gardeners who had been working at Kew for more than two years. The latter group, of course, included the Secretary of the Kew Employees Union, Will Purdom, who had by now been working at Kew for just under four years.

The Kew Union reacted swiftly, pointing out that the 1902 and 1903 intakes of student gardeners had entered Kew on the basis of the (then) new application form, which made no mention of a fixed term of employment. Accordingly, these Kew employees were subject to the same rules and regulations as the rest of the Civil Service and could only be dismissed in accordance with the Civil Service disciplinary code. The Union lobbied Labour MPs and the Board of Agriculture and Fisheries in this sense.

This legalistic argument from the Union was peculiarly difficult for the Board of Agriculture to resist because at the beginning of the year both the Board and HM Treasury had privately welcomed the appointment as Director at Kew of someone who was not a member of the Hooker dynasty, which had provided three successive Directors over the previous 64 years. The Hookers (who for these purposes included Thiselton-Dyer, the son-in-law of Sir Joseph Hooker) were considered to have run Kew as a private fiefdom, and Prain's appointment was welcomed as an opportunity to bring Kew firmly under the control of central government. Obviously, to affirm that the terms of service at Kew had been and remained distinct from those of the rest of the Civil Service would be inconsistent with achieving the strategic goal of folding Kew into the public administration on the same footing as any other government department. On the other hand, the arguments advanced by Prain, that student gardeners had always moved on after two years, or sometimes rather longer, and that such fixed terms were crucial to the functioning of Kew as a training establishment because until the departure of one cohort of student gardeners no new students could be taken on, were irrefutable.

Faced with this tangle of dilemmas, the Board reacted in classic Civil Service fashion, by refusing to make a decision. Prain was given to understand that this was a problem of Kew's own making for which the Director should come up with a solution acceptable to all concerned. Until then, no student gardeners should be discharged.

The ensuing discussions at Kew cannot have been easy. The ten student gardeners still at Kew who had entered service on the basis of the 1902 form stood on their rights as members of the Civil Service. Prain countered that they had never been in any doubt their employment at Kew was time-limited. He pointed out that whilst 21/- a week was just about sufficient for young men making their way in the world to live on for two years, it was quite inadequate as a long-term living wage. At the same time, he reminded the student gardeners that, since private individuals and colonial administrators frequently wrote to him asking if he could recommend someone for a particular post, the Director effectively had in his gift a number of better-paid positions in Britain and overseas. The student gardeners responded that gardening and horticultural positions at home were hardly ever filled in the autumn or winter because British employers preferred to save on the wages bill during the time of year when there was little to do in the garden: given the difficulties which even skilled men were experiencing in finding work they would not agree to being cast out from Kew until they had new posts to go to.

In November 1906, Prain advised the Board of Agriculture that agreement had been reached. The student gardeners who had joined on the 1902 application form would actively look for employment outside Kew (and, unstated but understood by both sides, Prain would be proactive in helping them find suitable positions), and as soon as they had secured new posts they would be discharged. Meanwhile, they would continue to be employed at Kew. The Board approved the arrangement, and it does not seem to have occurred to anyone (except, perhaps, Will Purdom) that it gave any of the student gardeners who had joined on the basis of the 1902 form and who were willing to live on 21/- a week an open-ended contract of employment at Kew.

It appears that these discussions were prioritised by the Kew Union over the question of student gardeners' wages, but by mid-1907 the Union and its supporters were once again able to focus on pay.

CHAPTER 5

The trap was sprung on 23 May 1907, when the Supply Committee of the House of Commons debated whether to approve the Board of Agriculture and Fisheries budget, including the salaries and expenses of the Royal Botanic Gardens, Kew. The Labour MP for Sutherland, Thomas Summerbell, proposed a reduction of £500 in the budget, to call attention to the position of the Kew gardeners. Summerbell contrasted the 21/- a week paid to the gardeners with the 27/- paid by the London County Council to gardeners in the Royal Parks, and opposed the decision to appoint a Deputy Director at Kew at a proposed salary of £500 a year.

On behalf of the Board of Agriculture and Fisheries, Sir Edward Strachey,* responded by saying that the student gardeners at Kew were

> *"not gardeners in the ordinary sense. They are really apprentices learning their work, and fitting themselves for good and remunerative employment elsewhere [...] They go to Kew to fit themselves for work in other parts of the country and the Empire. They merely receive their salary as apprentices; the 21/- is subsistence money."*

He went on to say that although 21/- would, if it were considered to be a weekly wage, be inadequate, and was less than the labourers at Kew received, *"these men [the student gardeners] are learners and the wage [sic] they receive is an increase on what they received some years ago"*.

Summerbell and his Labour colleague Joseph Duncan protested, Summerbell pointing out that the student gardeners had served five-year apprenticeships before coming to Kew and Duncan, who was living on an allowance of £2 a week from the Labour Party,† saying that he could personally confirm that 21/- a week was not a living wage. The vote was then taken and Summerbell's amendment was defeated.

Sir Edward Strachey's remarks in the debate were widely reported in the press. In the fiercely hierarchical society of Edwardian Britain it was a fundamental truth that a journeyman craftsman was a rung or two higher up the social ladder than an apprentice and that he was entitled to have this status recognised, including by his social superiors. Sir Edward was strongly criticised for describing skilled men as *"apprentices"*.

The Annual General Meeting of the Kew Guild on 27 May voted unanimously for a resolution proposed by William Dallimore,

> *"That this meeting protests against the statement made officially by Sir E. Strachey in the House of Commons [...] that the men employed at Kew are not gardeners but apprentices. They are men of the average age of 23 who have had at least five years' professional training before entering Kew as journeymen gardeners [...] and it is a misrepresentation of the facts to describe them as apprentices."*

---

* Sir Edward was not, as is sometimes said, the Secretary of the Board of Agriculture, but the Board's House of Commons spokesman: Lord Carrington, as a member of the House of Lords, could not speak in the Commons.
† MPs were not paid salaries until 1911.

The following day, a meeting of student gardeners at Kew resolved that
> *"we as gardeners and not as students decline to attend any lectures [...] in the Royal Botanic Gardens in lieu of wages. We desire a legitimate wage and shorter hours, leaving us free to attend lectures where we wish".*

In effect, the student gardeners were 'working to rule' to protest against the assertion that they were apprentices receiving subsistence rather than wages, as well as to support a demand for better pay and shorter hours.

The assertion on the part of the student gardeners that there was nothing in their terms of service requiring them to attend lectures was quite correct, which made it impossible for the Kew management to punish them for the boycott. This was a highly unusual situation for a British employer faced with strike action by a group of workers.* The Kew strike was, objectively, a minor industrial dispute, but it was important to Labour MPs and to trade unions as a rare instance of a trade union using strike action to bring pressure to bear on an employer without penalties being immediately imposed on the workers involved.

The Kew application form was promptly changed so that student gardeners joining after mid-1907 were required to attend lectures, but for the rest of the year not a single student gardener entered the lecture-hall, which was efficiently picketed by the Kew Union to discourage waverers. Meanwhile, Richmond Council began to run evening classes in botany specifically for the benefit of Kew student gardeners.

By the spring of 1908, the situation had evolved to Prain's advantage. The newly joined student gardeners attended lectures, as required by their contracts of employment. All but one of the '1902 form' intake had left Kew. Will was now the only student gardener who was in a position to argue that his contract of employment was open-ended.

Prain was delighted when, in October 1908, Charles Sprague Sargent, the Director of the Arnold Arboretum in Boston, approached him to enquire whether he could recommend a candidate for the post of Assistant Superintendent.

The Arnold Arboretum had been established just under 40 years previously by the trustees of the will of James Arnold and Harvard College. Under Sargent's direction and in close cooperation with the Biology Department of Harvard, 'the Arnold' had rapidly become a world leader in the field of forestry and silviculture. The post of Assistant Superintendent came with a house and a salary of $1,000 (a little more than £180) a year. The Superintendent was elderly and in poor health and Sargent made it clear that the new appointee could expect to be promoted to Superintendent in the not-too-distant future, whereupon his salary would double.†

---

* In December 1901, in the *Taff Vale Railway* case, the courts had ruled that trade unions and union officials were liable to pay damages to employers for losses sustained as a result of strike action. This decision was reversed in December 1906 by the Trade Disputes Act, which also legalised peaceful picketing. But it was still lawful to dismiss workers who had 'repudiated their contracts of employment' by going on strike and many employers considered this to be an entirely reasonable course of action.

† In the event, the Superintendent, Jackson Dawson, died in 1916, still working in the Arboretum.

CHAPTER 5

Sargent wanted to find a man who was both an expert on trees and a skilled propagator who could grow on the cuttings which the Arnold Arboretum received from all over the world, often in poor condition because they had been a long time in transit. Will, who had by now spent over five years as sub-foreman in the Kew Arboretum, possessed the knowledge of trees and silviculture Sargent was looking for and was also an exceptionally successful propagator. Prain had no hesitation in recommending him to Sargent, and Sargent promptly wrote to Will offering him the post, to start work in Boston on 1 January 1909.

Neither Sargent's letter nor Will's answer has survived, but Sargent wrote to Prain on 30 October that *"Purdom [...]*

Charles Sprague Sargent, Director of the Arnold Arboretum, Boston, a 1919 portrait by his cousin, John Singer Sargent.   Sargent House Museum

does not seem to be satisfied with the offer I made him". On 21 December, writing from Boston,[33] Sargent gave Prain more details: Will (whose salary at Kew was £84.35 per annum) had told Sargent that he would take the job for a salary equivalent to that paid to the Curator at Kew, *i.e.* £300 per annum and a free residence,[34] but not otherwise. Sargent told Prain

> *"I think he is making a mistake as one hundred and eighty pounds a year is a good deal better than anything he is likely to get in England at present. I shall write him again and possibly a word from you might help him to a better understanding of the situation."*

Prain did not speak to Will as suggested, partly because he believed that Will wouldn't listen to him, but mostly because he considered Will's rejection of Sargent's offer to be so perverse, so completely unreasonable, that he wanted nothing further to do with him.

It is impossible not to feel some sympathy for Prain. He had, by his lights, 'played fair' with Will. Despite what Prain saw as a sustained campaign of studied insolence on the part of the student gardeners, largely organised by Will, he had recommended Will for a position most of the Kew student gardeners would have been more than happy to accept, and Will had turned it down!

Of all the twists and turns in Will's life, this incident is the one which has cost me the most sleepless nights. Will was a perfect fit for the post he had been offered, which effectively guaranteed a rewarding career in a horticultural speciality in which he was particularly interested. The initial salary was over double what he was currently earning, and at least equal to what most of his peers amongst the student gardeners received on leaving Kew. There were good prospects for promotion. True, he would have to leave Britain for the east coast of America, and the cost of living in Boston was higher than in London, effectively devaluing the salary on offer by around ten per cent,[35] but he was well aware that the publication in the British press of his name as one of the leaders of the Kew strike meant that many potential employers considered him to be a troublemaker, not to be hired on any account.[36] What could possibly have justified his turning Sargent down?

A number of hypotheses present themselves. We could simply accept Will's assessment of his own worth at face value. He was not much inclined to modesty, and he may sincerely have believed he was worth £300 a year. But would that have justified his turning down £180 with the prospect of £360 in due course? Is it not more likely that for some reason he did not want to leave Britain at this time? It seems possible that the explanation for his disinclination to leave Britain may lie in his trade union and political activism.

Will's superior in the Kew Arboretum, William Dallimore, would certainly have known about Sargent's offer and Will's response. In his unpublished memoir he says that "*From little things he* [Will] *told me I gathered that his political friends urged him to remain as a test case, they would see that he did not suffer.*"[37]

At this time the Labour movement was in the early stages of transforming itself from a federation of trade unions with a small number of associated MPs into an alliance between the trade unions and a mainstream political party. It was a time of flux and growth for the new party, which needed urgently to recruit both members and organisers. Labour Party officials considered that Will had done an exemplary job as Secretary of the Kew Employees Union, to the point where in 1907 the Party asked him to act as mentor for a group of nurserymen who wished to start a Nursery Employees Union.[38] It may be that those same officials or Will's friends in the House of Commons, Will Crooks and Thomas Summerbell, held out to him the prospect of a post as a trade union organiser or even as a Labour parliamentary candidate in the next General Election. The vigour and determination with which Will had argued the Kew Employees Union's case suggest that he would have found the prospect of such a position very tempting. Dallimore's observations support the notion that this may be the explanation for Will's apparently perverse decision to refuse Sargent's offer.

Sargent was taken aback by Will's refusal, but Prain's reaction was cold fury. On 2 November he wrote a twelve-page note to Sir Edward Strachey in response to a letter from Summerbell asking that the decision to end Will's service at Kew in December 1908 be reconsidered "*until he has found a suitable position.*" Prain recited the history of the dispute about Will's service in detail, and said at length and with considerable emphasis that to grant the extension requested

CHAPTER 5

by Summerbell would mean *"the destruction of Kew as an Imperial centre of technical training."* The Director did not say in so many words that if the Board of Agriculture decided to extend Will's contract then he himself would resign, but this was clearly implied.

The Board had also had enough of Will, and on 9 December Edward Shine, one of the clerks at the Board, wrote to him saying that he was instructed by Lord Carrington to say that *"the Notice* [of dismissal] *was issued by his direction and he has no intention of cancelling it."*

Will now had just two weeks before the end of his term of service at Kew. Dallimore records that *"no help was forthcoming from those who had made him* [Will] *promises."*[39] His record as an 'agitator' would prejudice many potential employers against him. Overall, his prospects were bleak. But almost at once he received a quite unexpected offer.

Chapter 6
New directions

Charles Sprague Sargent had come to Britain in September 1908 to engage a plant-collector to travel to north-west China to collect plants and seeds for the Arnold Arboretum. Ernest Wilson, the collector whom Sargent had sent to China in January 1906, had made it clear that he would not extend his two-year contract, during which he had despatched thousands of herbarium specimens, seeds, bulbs and cuttings to the Arboretum and the other subscribers to the expedition.

Sargent had agreed in 1906 with the US Department of Agriculture that Wilson would work in partnership with Frank Meyer, the US Department of Agriculture collector in China. Meyer, whose main interest was in plants of economic value, would also collect ornamentals in northern China, and Wilson would collect useful plants for the Department in the southern zone. But Sargent was bitterly disappointed by how few ornamental specimens Meyer sent from Shansi (now Shanxi) province and was furious when these specimens were discovered to include several previously unknown species of larch, spruce, and pine from which Meyer, who had not recognised them as novelties, had not collected seed. Wilson, by contrast, was spectacularly successful, sending back thousands of herbarium specimens and large quantities of plant material, in the process enhancing the reputation of the Arboretum.

Sargent, a man of strong opinions and personal self-confidence verging on arrogance, refused to accept Meyer's explanation that the north of China was *"an utterly barren region"*[40] when it came to ornamentals and wanted to send a collector there to prove him wrong. Sargent also wanted this collector to harvest the botanical riches he was convinced were to be found in the high mountains of Kansu and Shaanxi (now Gansu and Shensi) provinces in north-west China.* He had come to England to persuade Harry Veitch to split the cost of a three-year expedition to north and north-west China, and to recruit George Forrest, the former Royal Botanic Garden Edinburgh collector, to lead it.

Harry Veitch was happy to agree to his old friend's proposal, but although Forrest came to London in September to meet Sargent and Veitch he refused their offer. Forrest was not impressed with Sargent's initial offer of an annual salary of £200, later increased to £300. He was reluctant to collect outside Yunnan

---

* Sargent believed that since they endured harsh winters in their home range, plants from north-western China would be better able to stand the New England and north European winters than those from further south. The logic is seductive, and such plants will indeed withstand bitterly cold winters, but they are very vulnerable to late spring frosts, having evolved in a climate where spring is a brief prelude to a hot summer, a short transition from extreme cold to baking heat.

CHAPTER 6

province, which he knew well and where he believed – quite correctly – that much more remained to be discovered. Finally, Forrest flatly refused to travel to China in January 1909, the deadline on which Sargent insisted, because he wanted to be at home for the birth of his first child in April. Sargent had to return to Boston in October, leaving Veitch to find a collector.

After three months during which Veitch failed to propose a candidate, Sargent wrote to him in early December reminding him of their agreement to send a collector in early 1909 and suggesting that Purdom might be a good man for the job. Veitch received Sargent's letter on 14 December and promptly went to Kew to talk to Prain and William Bean, the Director of the Kew Arboretum. Both Prain and Bean told Veitch that he was better off without Forrest,* and Dallimore concurred. All three praised Will's knowledge of trees and shrubs and agreed that he would be a good man to send to China, but they all doubted that he would be willing to go. They were, however, unable to suggest anyone else and agreed to ask Will to call on Veitch at his office in Chelsea on Friday 18 December.

At their meeting, Will told Harry Veitch that the suggestion to go to China had taken him by surprise: he had never thought about travelling overseas to collect plants, although he had been interested in reading about various collectors and in handling seeds sent to Kew from abroad, including from China. Veitch offered him a salary of £200 a year for a three-year expedition, plus £400 a year expenses and his passage to and from China. Will asked for a few days to think about it. Veitch agreed to this, and reported by letter to Sargent that *"on the whole I think Purdom was rather taken with the idea"*.

Harry Veitch's letter to Sargent crossed with one from Sargent to Prain. Sargent, a man who disliked taking 'no' for an answer, was unhappy that Will had turned down the Assistant Curator post and feared that this refusal reflected a lack of enterprise and an unwillingness to take risks. Did Prain think Will possessed the *"enterprise and grit"* needed in a plant-collector working in northern China?

Prain's reply to Sargent, dated 31 December 1908, the on-the-record character reference provided by Kew concerning Will, is a masterpiece of careful drafting. He told Sargent that *"we here have the highest opinion of his intelligence and his knowledge of plants and the opinion of his friends has always been that he was a pushing, energetic young man."* Prain had *"not experienced anything to indicate that he was deficient in enterprise and grit"* (true enough, but a considerable understatement!) and he had told Harry Veitch that he thought Will was well-fitted for the work.

At the very end of December, Will wrote to Harry Veitch accepting his proposal in principle but expressing some doubts about how he would cope in China, a

---

* It does not seem to have occurred to Veitch that against the background of the historic rivalry between the Kew and Edinburgh botanic gardens it was hardly likely that anyone at Kew would praise an Edinburgh collector.

country he knew almost nothing about. Harry Veitch was on good terms with Sir Robert Hart, who in April 1908 had retired* after 48 years of service as Inspector General of the (Chinese) Imperial Maritime Customs Service, and was now one of his neighbours in Chelsea.[41] Harry was also an old friend of a former medical adviser to the Customs Service and distinguished amateur botanist and plant-collector, Dr Augustine Henry. He immediately arranged for both men to brief Will.

These briefings were a real stroke of luck for Will. Sir Robert Hart had personally moulded the Chinese Imperial Customs Service into an efficient and incorruptible government agency, employing some 1,500 foreigners and ten times that number of Chinese people to provide a reliable income stream for the Chinese government. He had created the first national postal system and had made navigation along the Chinese coast immeasurably safer by building a network of lighthouses. Hart was universally recognised as having been a dedicated and effective servant of the Chinese government,† personally strongly committed to promoting the international image and interests of China. He was referred to in the highest circles of the Chinese government, where he had many personal friends, as 'our Hart'. Before leaving Peking, laden with honours, he had been received by the Emperor and the Empress Dowager to make his farewells.

In short, there was literally no one in the Western world who understood contemporary Chinese politics better than Sir Robert. The only possible caveat was that his personal perspective was based on unequivocal support for the Ch'ing government, which he had served for five decades, and especially for those Chinese politicians at the Ch'ing court who believed passionately that China needed urgently to master and adopt Western technology. It would be surprising if, in addition to advice about dealing with Chinese officialdom, Hart had not briefed Will about the need for China to improve its agricultural and forestry management, amongst other reasons in order to guarantee an adequate supply of timber for the construction of a national rail network.

Dr Augustine Henry, an old friend of both Sir Robert and Harry Veitch, had served for 19 years in the Imperial Customs Service as a medical officer. During this time, he had taught himself botany and had sent hundreds of live plants and seeds to Kew and James Veitch & Sons and over 15,000 herbarium specimens to Kew. The latter included leaves and flowers from *Davidia involucrata,* the dove tree, and when in 1899 Harry Veitch had sent Ernest Wilson to China, it was Henry who had drawn Wilson a sketch map showing where the tree was to be found. Wilson had returned with 15,000 seeds, and memories of the profit his firm had made on dove tree seedlings may well have influenced Harry Veitch's decision to sponsor Will.

---

* Technically, Sir Robert was on home leave, but all concerned understood that he would never return to China.
† This was the contemporary judgement, to the point that Hart was criticised by some British observers for having 'gone native'. Modern historians question why foreigners collected taxes in China, and why they paid themselves such handsome salaries for doing so.

CHAPTER 6

Henry, who had graduated in 1900 from the French National School of Forestry in Nancy, was engaged in writing, with botanist Henry John Elwes, the monumental *Trees of Britain and Ireland* for which he is principally remembered today. He travelled to London from his home in Cambridge in early January to give Will the benefit of his experience in recruiting and managing Chinese staff in the field. He is also likely to have discussed Chinese forestry and forestry management with Will.

Hart and Henry stood out from the majority of 'Old China Hands' in that both had gone to the trouble of learning to speak, read and write Mandarin Chinese to a good standard.* (Many foreigners didn't bother, preferring to rely on interpreters, even during sojourns in China extending over decades.) They both utterly rejected the casual racism with which the large majority of foreigners viewed, and treated, Chinese people. Will was a decent man with a strong sense of social justice who would have had no truck with such views even if he had not been briefed by Hart and Henry, but it seems unlikely to be a coincidence that on his arrival in China he immediately applied himself to learning Mandarin.

Will now returned to his former workplace, Coombe Wood, to familiarise himself with the plants Wilson had sent back from China in the last two years. He and Harry Veitch met there on 9 January. They agreed that Will would travel to China via Boston and Vancouver, and Veitch sent a triumphant telegram to Sargent: "*PURDOM WILL GO*".

The next three weeks passed in a blur, as different experts crammed Will's head with useful knowledge about China and Chinese plants. Kew-based photographer E. J. Wallis also taught Will to operate the Rolls-Royce of whole-plate cameras of the period, a Sanderson 'Tropical' model using 5×7 inch glass plates which, once exposed, would be sent to Wallis for processing.† The Sanderson camera was an elegant but complex machine which may have caused Will a twinge of apprehension as Wallis explained some of its more sophisticated functions, and he also learned to use a simpler Kodak 'pocket' camera which utilised rolls of silver-nitrate film. Meanwhile, Sir Harry drew on advice from Henry to put together and pack the other equipment and personal effects Will would need in the wilds of China over the next two or three years.

Somehow Will also found the time to attend a meeting of the Kew Union where he was presented with a watch and chain and a fountain pen, generous gifts from a small community of poorly paid men who wanted to make clear their esteem. He made a weekend visit to Westmorland to say goodbye to his parents. Last but not least, he agreed with a young woman who lived with her parents in north

---

* Hart, as Director-General of the ICS had made the speaking, and reading and writing, of Chinese by foreign members of staff an absolute condition of employment in the Service, and enforced the rule by a system of three-yearly examinations.
† The plates could not be developed in Peking, and both Ernest Wilson and Will sent exposed plates to London for processing. There were very few breakages in transit.

London, Annie Valentine Groombridge, that if and when he returned from China they would marry.[42]

Annie Valentine* was the 19-year-old daughter of Amos Groombridge, a former Kew student gardener who had left Kew in April 1889 to work at Kensington Gardens before becoming a park superintendent in Shoreditch. Annie was a student teacher and the same age as Will's youngest sister, Nell; the two young women may have met in the course of their training or through his older sister, Margaret, a secondary-school teacher with whom Nell shared lodgings in central London. What is certain is that Annie was a friend of the Purdom girls who more than once spent part of her summer holidays at Brathay Lodge, so that she and Will may well have known each other for several years before they became engaged. Will was an intensely private man in all things concerning his personal life and although his family and Annie's knew and approved of the engagement there was no public announcement and he did not mention it to Veitch or Sargent.

Veitch arranged letters of introduction to Sir Alexander Hosie, the British Consul General in Tientsin, the port city of Peking, and to the China Inland Missionary Society, as well as letters from Rothschilds Bank to their Chinese correspondents. He reported to Sargent that Will was *"throwing himself into the work in a very whole-hearted manner"* and that he was *"much better pleased at our prospects with him than I should have been with Forrest"*. He also reported without comment the advice he had received from the Kew Curator, William Watson, that Will was *"a splendid worker but if anything too energetic, and might want a little checking"*.

Will sailed on the *Oceanic* from Southampton to New York on 3 February, and reached Boston four days later. Unbeknownst to Will, Sargent – who may have noticed what Prain had left unsaid in his letter of 31 December – had reserved the right to veto his appointment if on meeting Will he considered him to be not up to the job.

Fortunately, Sargent immediately formed a favourable impression of Will, and he spent Will's second day in Boston writing an eight-page closely typed memorandum of 'guidance' about where, when, and what Will should collect in China.

Sargent told Will that in his first season's collecting he should, after being briefed by Ernest Wilson, proceed to Peking and thence 120 miles north to Jehol (modern Chengde) and the old Imperial hunting-ground at Weichang to the north of Jehol. In a characteristic display of wishful thinking, Sargent asserted that since Weichang *"has never been covered by a* [Western] *botanist, it is not impossible that you will find many interesting and possibly entirely new plants"*. Will was to leave Weichang in August so as to be in the Wutai mountain range,

---

* She was born on 14 February 1889, hence her second name.

CHAPTER 6

180 miles south-west of Peking in Shanxi Province, in mid-September, in time for the seed-drop of the pines and birches: obviously, Sargent especially desired seed from the new species of which Meyer had sent herbarium specimens. Once the seeds had been collected, which Sargent thought "*ought not to take very long*", he hoped that Will would return, via Peking, to Weichang – a round trip of around 600 miles – to gather seeds and herbarium specimens there.

The second year (1910) was to be spent entirely in the south-western part of Shanxi Province, where Will was to seek 'the wild tree peony'* before exploring the Peling mountain range, around 500 miles south-west of Peking and immediately north-east of the ancient former capital, Xi'an.

Finally, the third year (1911) was to be passed in Kansu Province, in the high mountains on the border with Tibet, over 1,000 miles from Peking.

All this was spelled out by Sargent with admirable clarity, and he was equally clear about the principal object of the expedition, which was "*to investigate botanically unexplored territory* [and] *to increase the knowledge of the woody and other plants of the* [Chinese] *Empire*". In pursuit of this goal, Sargent expected Will to dry six sets of all woody plants, including specimens of the same species which might occur in different regions, so as to show the extent of any variation. He also wanted Will to photograph "*as many trees as possible*", including their flowers and bark, and "*if time permits* […] *views of villages and other striking and interesting objects, as the world knows little of the appearance of those parts of China you are about to visit*".

These goals were not quite the same as those articulated by Harry Veitch, who had told Will "*the object of your mission* [is] *to collect seeds and plants of trees and shrubs, also any plants likely to have a commercial value, such as lilies*", but there was sufficient overlap that Will felt he could satisfy both his sponsors. He must also have welcomed Sargent's brief acknowledgement that it might be impracticable to complete the ambitious itinerary he had sketched out in three collecting seasons and that Will might need, in the light of local advice or experience, to change it.

Sargent had his legal adviser draw up a contract, which he and Will signed. This stipulated that "*all seeds of herbaceous, alpines and bulbous plants and all bulbs and other roots except those of woody plants*" collected by Will would be the property of the firm of James Veitch & Sons and would be sent directly to them from China. Plant material pertaining to woody plants would be divided equally between Veitch and the Arnold Arboretum.† Photographs and herbarium

---

* So-called tree peonies are shrubs with permanent woody stems, unlike herbaceous peonies which die back every winter. They have been grown as garden flowers in China for centuries, and there are hundreds of cultivars. Sargent was hoping Will would find the original wild form, which was reputed to grow on the Moutan-shan mountain.

† In practice, after he had started sending material to Boston and London it became clear to Will that the latter reliably arrived six weeks after leaving Peking, and the former frequently took twice as long. Accordingly, he sent much of the material to be divided between his sponsors to London, on the basis that it was more likely to arrive in good condition than material sent to Boston. Veitch would then divide it as appropriate and send the Arboretum their share (a two-week journey from London).

specimens would belong to the Arboretum, but Will was entitled to make and to retain a second set of herbarium specimens.* The Arboretum would pay his salary and expenses each January and July,† after which Veitch would reimburse one-half of the total sum involved.

Will spent a fortnight in Boston, mostly being taught how to prepare herbarium specimens. This involves pressing specimens of plants, including as appropriate the leaves, stems, flowers, fruit and seeds in blotting-paper (also called 'drying paper') before mounting the specimens on acid-free card with a note of the name of the plant, if known, the date and site of collection and any details which may be lost as a result of pressing and drying, such as colour or scent. Ideally, the specimens should be mounted in such a way as to give a good idea of the appearance of the plant in the wild. It is a lengthy and laborious process, not least because of the need to change the blotting-paper every couple of days until the plants are thoroughly dried out.

After his training in Boston, Will travelled by train to Vancouver, where on 24 February he boarded the *Empress of Japan* and sailed for China. He arrived in Shanghai on 16 March 1909.

---

* This was a fairly standard arrangement between collectors and their sponsors, and the sale of the duplicates on returning home typically earned the collectors a useful bonus.
† The salary went to Will's bank account in London, which was managed for him by his elder sister Margaret, and the expenses were sent to a bank in China.

Chapter 7
Inside the Heavenly Kingdom: Paradise Lost

Will disembarked in Shanghai at a moment when the Chinese state was in crisis. The moral and legal authority of the old political order had been destroyed by repeated humiliation at the hands of foreign powers including Britain, France, Russia and Japan. In the course of the previous half-century, up to 20 million people had died in violent uprisings seeking to overthrow the government or to break free from rule by Peking. Provincial authorities in large areas of the country only obeyed the will of central government to the extent that it suited them to do so. Incredible as it may seem to anyone looking at the superpower that is China today, 100 years ago there was a real possibility that China might break up as a unitary state and the fragments become a patchwork of Western colonies.

Although he did not know it, Will would spend most of the rest of his life in China. I believe that an outline grasp of the history of the country before his arrival is essential to an understanding of Will's life and work there, but readers who disagree may wish to skip to the end of this chapter.

For many centuries, China was ruled by an alliance between the land-owning and literate class, around one per cent of the population, whom modern Chinese historians call 'the gentry', and the numerically tiny aristocratic elite clustered around the Imperial court. The gentry had a direct interest in maintaining the established order and the state had a matching interest in maintaining the social and economic authority of the gentry.

In the 17th century, the rulers of the territory immediately to the north of China, Manchuria, overthrew the Ming emperor to establish the Ch'ing dynasty, commonly referred to in the West as the Manchu dynasty. Although the Manchu were committed to maintaining their separate ethnic identity, they co-opted the predominantly Han Chinese gentry, who competed in the annual Imperial examinations for jobs in the civil service and as army officers and who frequently acted as tax-farmers for the government.

The whole system was kept on track by adamantine family and clan-based social structures based on and enforcing a Confucian ethical/social code grounded in absolute respect for hierarchy and seniority and unquestioning compliance with ritual and precedent.* Punishment for those who rebelled, whether peasants

---

* For example, the adolescent Tung-chih (in pinyin, Tongxi) Emperor much resented the fact that if he tarried in a Peking brothel until after the gates to the Imperial palace complex, the Forbidden City, had been closed for the night (a not infrequent occurrence), then he was locked out until the following morning. But no matter what he said to the sentries on the walls of the Forbidden City, even the Emperor could not persuade them to break the rules by opening a gate to let him enter.

challenging the authority of the local gentry, or gentlefolk challenging that of the Emperor, was brutal and was routinely applied to whole families, clans, or even entire towns and cities.

Until the late 18th century, China kept foreigners and foreign states very much at arm's-length, reflecting the absolute conviction of Chinese people that China was the centre of the world, had nothing to learn from barbarian visitors, and would derive no benefit from dealing with them. This position became increasingly untenable as Western merchants began to trade with their Chinese coastal homologues. But the Chinese political and governmental establishment was not able – unlike, for example, the Meiji regime in Japan – to evolve a new philosophy to enable them to embrace the technology which had emerged from the Industrial Revolution in the West.

By the early years of the 19th century, then, Western merchants had established thriving trade links with China – but almost all in one direction. There was strong demand in the West for Chinese tea, silk and porcelain, but the only foreign goods that found a market in China were high-quality clocks, clockwork toys, ginseng from New England and the Appalachian mountains, and small quantities of woollen cloth and printed cotton. Chinese traders would accept nothing but silver in payment for their merchandise. By the 1820s China was running a substantial balance-of-payments surplus with Britain and America, the world price of silver was rising, and Western traders were having to pay more and more for Chinese goods. They began systematically to import into China large quantities of the one foreign commodity in demand there: opium.

Opium had long been grown in China for consumption in small quantities by the gentry, but large-scale Western imports of opium grown in Bengal made it much cheaper and thus accessible to all levels of Chinese society. The trade was illegal, but foreign merchants bribed Chinese officials, and by 1840, Lin Ze-Xu, the Emperor's special envoy, reported that the value of opium sales in China was running at over double the total budget of the Chinese government.

The Emperor approved Lin's plan to crack down hard on all those involved in the trade, and in March 1840 Lin arrested 1,600 Chinese citizens and confiscated from foreign firms nine million dollars' worth of opium, which was publicly destroyed.

British merchants demanded that their government force China to compensate them for their losses and to 'open up' the country to unrestricted foreign trade. The Royal Navy crushed Chinese resistance with humiliating ease to gain control of the Pearl River and the high ground overlooking the port of Canton. Hostilities were formally ended by the 1842 and 1844 Treaties of Nanking and Whampoa, by which China was forced to allow British merchants to import opium, ceded Hong Kong to Britain, opened five 'treaty ports' to foreign trade, and granted territorial 'concessions' where the writ of the Chinese government did not run, effectively British colonies within China. The Chinese government was also forced

INSIDE THE HEAVENLY KINGDOM: PARADISE LOST

to pay heavy 'reparations' for the losses sustained by British traders, funded by an international loan to be repaid from future tax income. This marked the start of several decades of systematic humiliation of the Ch'ing regime by Western powers, bullying which would become increasingly obvious to the people of China.

During the next half-century, there were repeated rebellions in China, but only two were truly national in extent – the Taiping and the Boxers. The Taiping rebellion, which began in 1850 in Guangxi in southern China, aimed to overthrow the Ch'ing dynasty and install a theocracy based on a much-modified and brutally authoritarian form of Christianity. It was initially very successful, and there were roughly simultaneous rebellions in the north of China (the Nian Revolt, when a poor and marginalised part of the country fought for a bigger share of national resources), and in the north-west (the Muslim Rebellion, also known as the Dungan Revolt, arising from a mix of religious and ethnic tensions), which diverted central government resources away from the struggle against the Taiping. By 1853, China had two capitals: Nanking,* the old Ming dynasty† capital from which the Taiping controlled the wealthy southern half of the country, and Peking, the capital of the increasingly beleaguered Ch'ing.

In 1858, by which time the government was starting to gain the upper hand in the civil war between the Ch'ing and the Taiping, a trivial incident in Canton (now Guangzhou) led British and French warships to bombard Canton, destroying both Chinese and British factories. In 1859 and 1860, the British and French governments mounted punitive expeditions which in 1860 captured Tientsin, the port city east of Peking. Emperor Xianfeng fled to Jehol, 140 miles from Peking.

In October 1860 French and British troops looted and burned the Old Summer Palace,‡ a complex of parks and buildings covering 850 acres north of Peking, before overrunning Peking. The Chinese government was forced to sign a new treaty paying compensation for the British factories burned in Canton, designating eight more treaty ports and giving Western countries the right to open diplomatic missions in Peking. The Emperor only made this last concession when told that the alternative was the destruction by British artillery of the walls of Peking, after which British and French troops would pillage the city, as they had Tientsin. Humiliated and ashamed, Xianfeng died at Jehol in August 1861.

The five-year-old heir, Hsien Feng, was proclaimed with the regal name Emperor Tung-chih (in pinyin, Tongxi). His mother, Ci-An, and Cixi,§ the former

---

* Re-named Tianjing (in pinyin, Tienking) 'Heavenly Capital' by the Taiping.
† The Ming dynasty preceded the Ch'ing.
‡ The echoes of the explosive charges with which British Royal Engineers destroyed the buildings of the Summer Palace reverberate in China to this day. The buildings and the surrounding park were the most beautiful of their kind in China and housed the greater part of the Imperial art collections built up over centuries. The soldiers destroyed what they could not carry away, but many iconic Chinese cultural treasures were shipped to Paris, London, or Berlin. A tiny Pekingese 'sleeve dog' from the Summer Palace was presented to Queen Victoria. She named him 'Looty'.
§ In the Wade-Giles transcription, Tsu-An and Tsu-Hsi.

concubine of Xianfeng, became joint Empress Dowagers. The appointment of two women as *de facto* rulers was unprecedented, but the Empress Dowager Cixi concentrated power in her own hands, brooking no challenge to her authority. She immediately gave the army leader Tseng Kuo-Fan (in pinyin, Zheng Guofang), who had risen from the ranks of the minor gentry after winning a series of decisive victories over the Taiping, near-plenipotential powers to prosecute the civil war.

This proved a shrewd move: Tseng smashed the Taiping and in 1864 captured their 'Heavenly Capital', Nanking. He then devastated a huge swathe of central China by ruthless ethnic cleansing of regions which had supported the Taiping. In 1868, another General, Tso Zongtang (in pinyin, Zhuo Zongtang), defeated the Nian rebellion before methodically and mercilessly destroying the Muslim forces behind the Dungan Revolt and systematically killing or driving out from 200,000 square miles of western China anyone who might be hostile to the Ch'ing regime. Then, in 1877, Tso led a successful campaign to throw the Russian-backed rebel Yakub Beg out of Kashgar and Xinjiang.

The immediate threats to the Ching dynasty were eliminated, but the price of victory was the creation of the first 'Chinese warlords', powerful political figures whose influence derived, not from their place in the Manchu hierarchy – Tseng, for example, was ethnically Han Chinese – but from the personal loyalty they commanded from an effective battle-hardened army. Both Tso and Tseng well understood the risks inherent in being perceived by Cixi as a potential threat and were at pains to display exemplary loyalty, but in later years other warlords would take quite a different line.

The Tongxi Emperor grew into an arrogant and wantonly promiscuous teenager who was denied any substantive power by the Dowagers, even after he had attained his majority. He died in December 1874, leaving no heir. His mother Ci-An formally adopted her three-year-old nephew, who became Emperor Kwang-Hsu (in pinyin, Guangxu). Ci-An and Cixi continued as Empress Dowagers: in reality, nothing changed.

The Nanking Treaties set the pattern for a series of treaties which the Chinese government was forced to sign in the second half of the 19th century with, amongst others, the governments of Britain, France, Russia, Belgium, Japan and America. By 1905, there were 48 'treaty ports' and thousands of square miles of China were under the rule of foreign governments.

Over the same period those same governments repeatedly exacted from China very large reparations for alleged treaty breaches and for indignities suffered by their citizens. Despite substantial tax increases and the creation of an onerous new tax on the movement of goods (from which foreign firms were exempt), the interest payments on the foreign loans from which this compensation was paid consumed half the income of central government. Once the expenses of the court and the pensions to which senior Manchu nobles were entitled had been paid, there was little left for public works such as maintaining river embankments. One consequence of this was the growing reluctance of Provincial authorities to remit taxes to Peking, since they received little in return. This further worsened the government's financial position and made it even harder for the central government to earn the support of Provincial authorities by addressing local needs.

Despite the deaths of up to 20 million people in the Chinese internal conflicts of the 19th century, by 1900 the population of China had grown to over 400 million. Unemployment and pressure on agricultural land increased, and the Peking government's incompetence and evident subservience to foreign governments reduced respect for the Imperial Throne amongst both the gentry and the peasants. Local militias, even local armies, proliferated.

Amongst the concessions insisted on by the Western powers in the 'unequal treaties' was the right of Christian missionaries to go where they pleased in China and for all practical purposes to be above Chinese law.[*] Missionaries fanned out all over China and were hugely disruptive of social structures in the provinces, not least because many of them, especially the Roman Catholics, demanded that the legal immunity they enjoyed should be extended to Chinese Christians. Some Chinese people converted for personal advantage, for example to win lawsuits or to justify refusing to pay their share of communal charges to maintain temples. This stoked anti-foreigner sentiment in the countryside, where a series of natural disasters was also blamed on the presence of missionaries.

In the mid-1890s the 'Boxer' rebellion (so called because its followers claimed to commune with the Gods through the medium of traditional Chinese boxing) arose in Shandong Province, immediately south of Peking. The slogan and political agenda of this violently anti-foreigner movement was 'Defend the Ch'ing, destroy the Foreign', so it can be argued that it was not, strictly speaking, a rebellion. But it was highly disruptive of law and order across the nation and, whatever its adherents said to the contrary, seriously challenged the authority of the Ch'ing regime.

The credibility of the government was further reduced by Britain overrunning Burma, and France overrunning the Annam kingdoms (now Vietnam). Japan did much the same in Korea, and Japanese forces defeated with humiliating ease the supposedly modernised Chinese armed forces that opposed them. None of the three conquered nations was an integral part of China, but they were tributary states whose rulers recognised, at least in theory, Chinese sovereignty. Their loss further eroded the authority of the Ch'ing government, and led many Chinese people to fear that the next move of the Western powers[†] might be to colonise parts of China itself.

---

[*] In theory, missionaries could be prosecuted for criminal acts or have civil actions brought against them in consular courts in the nearest concessions of their home countries. In practice, the former never happened, and the latter was a waste of time.
[†] Including, for these purposes, Japan.

CHAPTER 7

In the spring of 1900, the Boxers, by now a national movement with millions of adherents, and their allies, including a violent vegetarian sect dedicated to killing meat-eaters, murdered nearly two hundred foreigners and tens of thousands of Chinese converts before marching to Peking, where they laid siege to the Legation Quarter in Peking.

The besieging forces comprised Boxer troops, who fought bravely but whose use of spears and swords rather than firearms and reliance on magic charms to protect them against bullets made them ineffective, and government soldiers. The main contingent of the government forces supporting the Boxers largely refrained from using artillery and usually fired high when they did. They also made copious use of fire-crackers as an alternative to rifle-fire. Their commander, Jung-Lu (in pinyin, Rong-Lu) understood that whilst his forces could have overrun the Legation Quarter, the Boxers would then have killed all the 924 foreigners and the 3,000 or so Chinese Christians within the perimeter, and foreign governments would almost certainly have responded by overthrowing the Ch'ing regime and partitioning China into Western colonies.

Despite Jung-Lu's efforts to keep casualties amongst the besieged as low as was consistent with being able to maintain to the Dowager Empress that he was genuinely trying to capture the Legation Quarter, over 60 foreigners and several hundred Chinese Christians were killed in the siege. Thousands of Boxers and hundreds of government troops also died.

Empress Dowager Cixi, who hoped to use the Boxers to change the balance of power between foreign states and China, declared war on eight Western powers on 16 June. This was a catastrophic error: in September, the siege of the legations was lifted after 55 days by a 15,000-strong expeditionary column of troops from the 'Eight Power Alliance'. The Alliance then occupied and looted Peking. Cixi and the court fled to Xi'an, the ancient former capital of China, from where the government declared the Boxers outlaws and appointed the elderly general Li Hung-Chang (in pinyin, Li Hongzang), who was known to have opposed Cixi's cooperation with the Boxers, to negotiate a peace settlement.

Li did his best, but the settlement terms were harsh; the payment of an indemnity (funded by yet another international loan) exceeding the entire annual income of the government and further consolidation of the treaty ports and concessions system. The silver lining was that Western powers were now owed so much money by the Ch'ing regime that they had a vested interest in its survival.

Another unexpected benefit of the settlement was that the wholesale purge of reactionary officials on which the Western powers insisted as part of the settlement gave the 'modernisers' at court a relatively free hand to push for change, including reform of the civil service and the adoption of modern contrivances such as river steamers. An ambitious railway network was planned and began to be built. Modern schools, teaching foreign science and history, were established, including some for girls. The dreadful cruelty of foot-binding for Han women* was not quite outlawed, but was strongly discouraged. The ban on intermarriage between Han and Manchu was lifted, and there was a crackdown on the cultivation and use of opium.

Historians disagree about whether Cixi, who had three years earlier imprisoned the Emperor and executed six of his associates for espousing similar ideas,† experienced a Damascene conversion after seeing with her own eyes during her journey from Peking to Xi'an the suffering of the rural population, or whether her authority was so compromised as a result of her error in supporting the Boxers that she had no alternative but to allow the modernisers to proceed. In any event, she continued to keep the Emperor closely confined on an island in a lake in the palace grounds.

The modernisation programme triggered something of an economic boom as the more populous regions of China began to experience an industrial revolution. But the tax increases required to pay for it, which bore particularly hard on the rural poor, were widely resented, and the movement of people and resources from rural areas to fast-growing cities was disruptive of established social and commercial norms. Inflation, especially the cost of basic foodstuffs, soared. The Han Chinese population became increasingly unwilling to accept Manchu economic hegemony. This led to inter-communal violence and further weakened the links between the Han gentry and the government.

In November 1908, the Empress Dowager and Emperor Kwang-Hsu died within 30 hours of each other. She was 73 years old and had been ill for some time; he was 38. He had never enjoyed robust health, but the illness which carried him off was sudden and brief. Poisoned, many believed, and modern forensic science has proved them right.‡ Cixi survived the Emperor just long enough to forge his last will and testament, nominating his successor, the not-quite-three-year-old Pu Yi, and recording the late Emperor's alleged dying wish that the Imperial edicts published in September 1906 and August 1908 proposing the creation of elected regional assemblies and of a national Parliament composed of representatives of those assemblies should be implemented.§ The will also appointed the arch-conservative Prince Chun as Regent, which infuriated the modernisers at court and in the short term made it impossible for the government to reach agreement on any meaningful policies.

---

* To understand foot-binding, make a fist with your hand with the thumb outside and imagine that the bones in your foot have been smashed with a wooden club, folded over and tightly bound so you have to walk on the knuckles of your four smaller toes, which are tucked into the soles of the feet, like a fist.
† To be strictly accurate, at least eight of the Guangxu Emperor's associates were members of a group of modernisers which in 1898 planned radically to modernise Chinese society, the economy, and government. Several amongst the group believed that removing Cixi was essential if reforms were to be achieved, and plotted to do so. Cixi had four of the plotters beheaded, along with two other modernisers who were not parties to the conspiracy. Two more conspirators fled into exile.
‡ A detailed analysis of the Emperor's mortal remains proving that he had died after ingesting a single large dose of arsenic was published in the *Beijing News* on the centenary of his death, 3 November 2008.
§ In truth, it is likely that the Emperor felt these proposals did not go nearly far enough towards establishing democratic institutions with substantive executive powers.

CHAPTER 7

Thanks to the briefings he had received from Augustine Henry and Robert Hart, Will was familiar, at least in broad outline, with these events. He would also have followed at the time the accounts in the British press of the campaign against the Boxers, and read the despatches of the *Times* correspondent in Peking, George Ernest Morrison, which presented a detailed picture of current events in China.*

Despite the Cambridge briefings, however, it is unlikely that either Will or Harry Veitch fully understood the considerable risk run by foreigners who ventured into the remote provinces for which Will was heading. Charles Sprague Sargent had been specifically warned about this: when Ernest Wilson heard of Cixi's death and that of the Emperor, he had written to Sargent from his camp on the Dadu river in Szechuan Province saying that

> "*jealousies carefully hidden whilst the Empress ruled will now hark out and it is hard to prophesy what may happen [...] I think China will be a very good country to be out of until the new Emperor is firmly established on the throne.*"[43]

But Sargent notoriously disliked taking advice or changing his mind once he had decided on a course of action.

Ernest Wilson had repeatedly made it clear that he would hold Sargent to their two-year contract and was not interested in extending it, but he was nonetheless aggrieved when Sargent wrote to say that Veitch and Sargent had engaged Will and hoped that Wilson would brief him before returning to London. Sargent seems to have anticipated Wilson's reaction, because he told him that Harry Veitch had refused to sponsor half of Wilson's salary of £750 a year,† and that this was the sole reason Sargent had not offered Wilson a new contract to continue collecting in China. Wilson replied on 9 March from Shanghai, where he was arranging for the previous year's collections to be shipped to Boston, expressing "*a slight feeling of chagrin at being passed over so completely in favour of another*" but promising that he would do anything he could to help "*your new man*".

In Shanghai, Will stayed with fellow-Kewite Donald MacGregor and his wife. MacGregor had been a student gardener at Kew between January 1902 and 1904, when he was appointed Supervisor of the Shanghai Parks and Gardens. Will's baggage had been consigned to MacGregor and he and Ernest Wilson, whom Will met on 25 March, gave Will useful advice on living and working in China.‡

---

* Some of those despatches, including the report of Cixi's death, were written by Morrison's part-time assistant, the baronet Sir Edmund Backhouse. Backhouse, a compulsive liar, fantasist, forger and pathologically misogynist pornographer, almost single-handedly created a legend of Cixi as a sexually voracious and depraved she-devil which to an extent persists today.
† It is unlikely that Veitch would have agreed to pay £375 a year as his half-share of Wilson's salary, but Sargent had not actually asked him! What was certainly true, though, was that Will's salary of £200 a year was less than a third of what Sargent was paying Wilson.
‡ There was another Kewite in Shanghai at this time, John Giles, who had been a student gardener from January 1904 until his appointment in August 1907 as Assistant Supervisor to the Shanghai Parks. No doubt Will also met Giles, but there is no surviving record of his having done so.

Ernest Wilson and Will may have met during 1905, when they were both working at Kew,* but Will had not been able to attend the lecture "*Wanderings in China*" which Wilson gave to the student gardeners on 6 November 1905,[44] because two days previously he had been dismissed by Thiselton-Dyer. Will had regretted missing the lecture, and was pleased that Sargent had been able to arrange for him to meet Wilson. Wilson, for his part, found Will a good deal more congenial than the Dutch-American plant-collector Frank Meyer to whom he had given a similar briefing, also in Shanghai, two years previously.

What is striking, though, is that Wilson did not propose that it would be to Will's advantage to engage any of a dozen or so Chinese men who had supported him over the last three years, and whose term of employment with Wilson would end as soon as they had finished packing the harvest of the last season's collecting for shipment to Sargent. If Will had hired some or all of them, he would have benefitted from their experience and their training in, for example, preparing herbarium specimens. The men themselves would surely have welcomed continuation of their employment. Wilson's reticence is all the more remarkable when one recalls that immediately after being briefed by Augustine Henry at the start of his first (1899–1902) expedition he had hired several of the assistants who had accompanied Henry on a six-month plant-hunting expedition in 1898. But Will lacked the experience to suggest he might do the same thing, and despite his promise to Sargent that he would do all he could to help Will, Wilson did not suggest it.

Wilson may have kept silent because he anticipated that he might return to China within the three-year period for which Will was contracted to collect for Sargent and Veitch. (In the event, in June 1910 Wilson did return, and promptly reconstituted his team of helpers. Obviously, this would have been impossible if the men had been in the field with Will.) A less charitable explanation is that Wilson wanted to nip in the bud any possible challenge to his burgeoning reputation as the greatest of the Western plant-hunters active in China. Certainly, in later years Wilson quite deliberately burnished his legend, including by re-writing some of the history of his first two expeditions.[45]

By chance, Sir Alexander Hosie, the British Consul General to Tientsin and *de facto* head of the commercial department of the British Legation, was visiting Shanghai at this time. Hosie had spent his entire diplomatic career in China and had lived in or visited almost every Chinese province. He was also a self-taught botanist of note who had sent thousands of herbarium specimens to Kew. Colonel Prain had, via Harry Veitch, provided Will with a letter of introduction, and Will also met Hosie on 25 March. Hosie told him that he was leaving Shanghai the next day for Tientsin, the port of Peking, where he would be spending a few days. As Will was catching a boat to Tientsin on the 27th, arriving on the 31st, perhaps they might travel together on the train from Tientsin to Peking so as to be able to continue their conversation? Will was happy to agree.

---

* Wilson worked at Kew Gardens between March and December 1905 and continued to live nearby for the whole of 1906. He greatly enjoyed his time at Kew, visited whenever he could, and was a frequent contributor to the *Journal of the Kew Guild*, of which he was a life member.

Chapter 8
Peking, and the 1909 plant-hunting season

Will arrived in Tientsin on 31 March and travelled the 80 miles between the port and the capital, in the company of Sir Alexander, who had a great deal to tell him about travel in the more remote parts of China.[46] Will mentioned that he also had a letter of introduction to the British Minister,* Sir John Jordan. Hosie promised he would mention Will to Sir John, one of his oldest friends, and would also introduce Will to some of the Legation staff.

Will had gone ashore when the ship bringing him to China had briefly docked in Yokohama and he had spent ten days in Shanghai as the guest of Donald MacGregor. But both Yokohama and Shanghai were cosmopolitan cities with large expatriate communities and commercial quarters where the buildings and the street scene were not wholly unlike the streets of London or San Francisco. Will's first sight of the massive crenellated walls of Peking, over 20 miles around and up to 40 feet high, dominating the featureless plain between Tientsin and Peking must have made it quite clear how very different was the world he was now entering.

Peking was an ancient Chinese city whose history was proudly proclaimed in its walls and streets. In addition to the main city wall, the 'Forbidden City', the Imperial Palace complex, half a mile square, was surrounded by its own tall walls. North of the Forbidden City was the area foreigners called 'the Tartar City', reserved for ethnic Manchus. South of this was the 'Chinese City', ten square miles of narrow streets and alleys crammed with the less prosperous inhabitants. The total population was of the order of 800,000 people, divided more or less equally between Manchu and Han Chinese.[47] Foreigners were a tiny minority, around 1,000 people, rather more than half of whom were permanent residents.

The great majority of the streets of Peking had changed not at all in the four centuries since they had been described by Marco Polo. Arrow-straight and running due north–south and east–west, they were crowded with blue-hooded Peking mule-carts, whose nail-studded wooden wheels made an incessant clatter on the stony road-surface, trains of soft-footed Bactrian camels laden with coal, wheelbarrows, donkeys, ponies and sedan chairs, street-hawkers, fortune-tellers and beggars. They were covered with animal droppings and human excrement, which meant that pedestrians had to be careful where they put their feet and

---

* As Ambassadors represent sovereign states, and the Ch'ing considered the Emperor to be sovereign over the whole world, they insisted that representatives of foreign governments in Peking could only be classed as 'Ministers'.

CHAPTER 8

made the whole city stink appallingly. The only signs of foreign influence were the very rare motor-cars,* and the ubiquitous rickshaws, introduced from Japan in the 1880s. Will was a confident and resourceful individual, but he must have been glad to arrive at the Peking railway station in the company of someone familiar with the city.

Hosie helped Will find his way to the 'Grand Hôtel des Wagons-Lits' very close to the British legation, where Will checked in. Hosie himself lived in the Legation Quarter, a gated community covering 300 acres and containing both the foreign Missions and the homes of most of the foreign residents. It was the model of a manicured European city suburb, although the 'Jade Canal' which ran through the middle of it also carried the sewage of the Forbidden City: in the summer especially, the stench was overpowering.

The following day, 1 April, Will called at the British and American legations and presented the letters of introduction procured by Harry Veitch and Charles Sprague Sargent, respectively. At the American Legation he had a good meeting with the US Minister, William Rockhill,[48] one of the tiny handful of Westerners who spoke the Tibetan language fluently and the author of *Land of the Lamas*, an account of his remarkable journey, 20 years previously, through north-western China and into Tibet. Rockhill promised to support Will's application for an internal Imperial passport to enable him to travel to Jehol and they discussed Will's plans to collect in Kansu and the borders of Tibet during the 1911 season.

Sir John Jordan promptly put in hand the application to the Foreign Ministry for Will's passport, but warned him that the Chinese administration was preoccupied with arrangements for the funeral procession of the late Emperor, to be held on 1 May, and that the document might take some time to come through. When Will expressed concern at the financial implications of the delay – hotels in Peking had taken advantage of the flood of VIPs in town for the Imperial funeral to double their rates – Sir John told him that he was welcome to stay in the bachelor quarters provided for Legation trainee interpreters and clerks, an offer he gratefully accepted.[49]

As well as being cheap, Will's new lodgings meant he was folded into the collegiate community of Foreign Office trainee interpreters, young men in their mid-twenties, which must have helped him to find his feet in Peking. It also made it easy for him to start lessons in Mandarin, which he did immediately. To his surprise, Will found the language came easily to him. He had since childhood been a gifted mimic, with a good ear for tone and accent, and he picked up the language very rapidly (in time, to their amusement, he even mastered the slang

---

* In 1907 a French magazine organised the world's first motor race, from Peking to Paris. Five cars and a motorised tricycle were shipped to Peking by would-be competitors. The Chinese government initially refused to allow the first part of the race to go ahead, but the huge crowds attracted by the unprecedented spectacle of motor-cars driving around Peking were so disruptive of life in the city that the authorities relented, simply to be rid of them. Remarkably, all five cars (but not the tricycle) made it to Paris.

used by muleteers and carters and the chaffing argot of inn-keepers).[50] He was also diligent in learning to read and write Chinese characters.

A rite of passage for foreigners newly arrived in Peking was the acquisition of a Chinese name composed of three characters, the first representing the family name and the second and third the two given names. Ideally, the three characters together constituted a phonetic rendering of the person's family name in his native tongue. Will's teacher, after listening carefully to his pupil's surname, suggested that this could best be rendered by the three characters 'Pao' ('Wave', which thus became Will's Chinese family name), 'Er' and 'Deng', meaning, approximately, 'wave that rises' or, as the tutor more poetically put it, "wave rising to infinity".[51] As was standard practice, Will commissioned a wooden ink-stamp or 'chop' of these characters, which he could use to print out visiting-cards and sign official documents.

In addition to the letter to Sir John Jordan, Harry Veitch had provided Will with a letter of introduction to Colonel John Abbott Anderson, who was attached to the British Legation. Anderson helped Will hire a personal servant and procure those necessities for travel in the wild which he did not already possess. He then invited Will to accompany him on a trip to Huailai, 100 miles north-west of Peking,[52] to give him the chance to learn how to deal with the challenges presented by travel in rural China.[53]

These challenges were formidable. There were no large-scale maps and travellers had to rely on their carters' or muleteers' knowledge of the road, with some help from a compass to ensure they were going in roughly the right direction. The highways themselves were appalling, narrow, unpaved, and often hugely potholed. Although the climate in spring and autumn was reasonably clement, rain turned the roads into quagmires during those seasons and in summer they were dusty and baked into deep ruts. No one travelled in winter if they could avoid it. Twenty to thirty miles was considered a good day's journey.

The best inns – the ones in the cities and towns – were dark, verminous and crowded, but at least they were built of brick and tile. Rural inns were often little more than mud-brick and thatch hovels where travellers were lucky if they did not have to share accommodation with their pack-animals,* and boasted even greater numbers of fleas, lice and cockroaches, not to mention fearless rats.[54] The only form of heating was the *k'ang*, a hollow brick bed-platform warmed by burning charcoal within it, ensuring that sleeping travellers were toasted on one side and frozen on the other.

Travellers cooked their own food, using provisions they carried themselves or purchased from local markets. (Will is also known to have shot game birds as

---

* At least there was usually a choice of lodgings, even in remote parts of the country: rural inns nearly always came in pairs, a short distance apart, because mules and Bactrian camels, the other beasts of burden in northern China, would *not* share a stable.

CHAPTER 8

the opportunity arose, and perhaps the occasional hare or small deer.) The range of produce on sale in any given area varied greatly according to the season but broadly speaking in northern and north-western China the staples were cereals, dumplings and noodles, eaten with mutton, and in the south rice and pork. In Tibet, the staple food was patties made from *tsampa,* flour made from roasted barley, and tea seasoned with lumps of yak butter. It was essential to boil all drinking-water.

Finally, although Western travellers were usually reasonably safe in towns and villages, where everyone knew that if harm came to foreigners the Peking government would severely punish both local officials and the wider community, away from the settlements, especially in remote areas, there was no shortage of bandits on the look-out for opportunities to rob passers-by. These gangs, however, were usually armed with spears and swords, or perhaps matchlocks, and rarely attacked travellers carrying rifles or revolvers.

Colonel Anderson's assistance was not entirely disinterested, and the trip to Huailai was not undertaken purely to get away from the dust-storms which made life in Peking thoroughly unpleasant in April and May, when strong winds blew thick clouds of fine sand off the Gobi Desert into the city.

Anderson's official title was Commandant of the Legation Guard, which comprised 30 men and two NCOs. This is not a credible level of responsibility for an officer of his rank: in 1917, the guard force was exactly the same size and was commanded by a Captain. Two military officers from the China Intelligence Section of the British Army in India are known to have been attached to the Legation staff,[55] and it is very likely that Anderson was one of them.

The British Secret Bureau, which in due course evolved into the Secret Intelligence Service, also known as MI6, was established in late 1909 in response to concerns about German military and naval expansion and German espionage activity. The mandate of the Bureau was tightly defined and any activity in China or Tibet was expressly prohibited. This was not due to any lack of demand for political and military intelligence concerning these two countries. On the contrary, the Indian colonial administration, the British Army in India, the Imperial Defence Staff and the War Office were keenly interested in the state of the countries to the north and east of India and especially in any Russian activities in the region. But all these bodies had their own intelligence staff, and most of them had, at least some of the time, officers collecting information 'in the field'. References in this book to the British secret services in the plural reflect this situation.

There were surprisingly few 'turf wars' between the different British secret services, partly because they tended to specialise in different geographical or, sometimes, functional, areas.* The China Intelligence Section of the (British) Indian

---

* For example, Naval Intelligence had a monopoly on British code-breaking.

Army was well-resourced and effective. It is likely that the other secret agencies of the British government recognised its primacy in China, although it would also have been the case that some officers in the British Legation corresponded with the intelligence staff of the Services of which they were members; for example, the Naval Attaché maintained contact with Naval Intelligence.

Anderson's job was to provide the Indian Army General Staff and the Imperial Defence Staff with up-to-date intelligence about the state of the armed forces in China including current troop deployments and capabilities, and also information on the state of the Chinese transport network (broadly defined). The Staff wanted information on everything going on to the north and east of India and were especially anxious to obtain maps that would help identify possible routes for a putative threatened Russian invasion of parts of China or Tibet, which might then serve as jumping-off points for a move against India. Maps of the Chinese border regions would also come in useful if China were to break up, when Britain would be competing with other Western powers rapidly to occupy as much Chinese territory as possible.

The Chinese government was well aware that the latter eventuality was a regular topic of conversation in Western Missions in Peking and surveying, particularly in the border regions, was strictly forbidden. But whilst a theodolite*, the standard surveyor's instrument, is almost impossible to disguise, a reasonably accurate sketch map can be made with the aid of a prismatic compass, which even the Chinese authorities had to admit was a sensible aid to travel in uncharted country. The area where Will was going to collect in 1909 was familiar to Anderson, but he knew that Delhi and London would be very happy to receive maps of Shanxi province, where Will was bound in 1910, and would consider maps of Kansu – on the main route to Tibet, largely unmapped – to be of even greater value. When he arrived in Peking, Will knew nothing at all about cartography, but it is a safe bet that by the time he returned from his trip to Huailai with Anderson he had acquired a working knowledge of basic map-making and, if he did not already possess them, a prismatic compass and drawing-board.†

On 10 April, Will celebrated his 29th birthday. The contrast between his situation as he entered his third decade and the circumstances in which he had celebrated his previous birthday was marked. On his previous birthday, he had been living frugally in a boarding-house in Kew, on a wage of £7 a month, a sub-foreman at Kew with little discretion about how and when to go about his work and with no obvious prospect for improvement in his pay or his professional prospects. A year later, he was master of his own destiny when it came to where, when

---
* Five years later, the plant-hunter and amateur cartographer Frank Kingdon-Ward wrote from Yunnan Province to the Royal Geographical Society in London complaining that he could only use his theodolite in secret, at night.
† Will probably already owned a compass, but it is unlikely he had a drawing-board: it is clear from the one and only sketch in his papers that he had no talent whatever as a draughtsman.

CHAPTER 8

and how to carry out his plant-hunting commission, his salary had more than doubled, and his day-to-day expenses were covered by an expense allowance.

The change in Will's social status was at least as great as the change in his financial position. In today's world, one of the few communities which mirrors the adamantine hierarchy of Edwardian society is the armed forces, and a reasonable analogy is that by signing-on as a plant-collector with Sargent and Veitch, Will had quit the sergeant's mess for that of the officers. In 1908, Will could never have shared a railway compartment with Sir Alexander Hosie, as he had on the journey from Tientsin, when he and Sir Alexander had chatted about botany and about the state of Peking and of China generally. Quite apart from the fact that in Britain Hosie would have been travelling in a first-class carriage and Will in Second or Third Class, even if they had somehow been thrown together Hosie would have been unlikely to engage in conversation with someone whom he would have seen at a glance was several rungs below him on the social ladder, an impression which would have been instantly confirmed on his finding out how Will earned his living.

Will took this self-portrait in Peking in 1909.
© British Library Board

It is apparent from photographs taken of Will around this time[*] that he had invested some of his new salary in his wardrobe and that he was enjoying wearing well-cut suits and highly polished shoes, the uniform of the male British middle and upper-middle classes. Expatriate communities are almost invariably more inclusive and more welcoming of new faces than society 'back home' and Will's confidence was boosted by his becoming aware that he was not by any means the only working-man-made-good to be accepted both by senior members of the foreign community and by Chinese officials as their social equal. He must have felt a real sense of achievement that at last he was 'on his way up in the world'.

At the same time, Will must have been shocked at how very different life in Peking was from anything he had previously experienced, including the ten days

---

[*] Most of these photos were taken *by* Will, using a clockwork timer on the Sanderson camera: he had an undeniable weakness for self-portraits.

he had spent in Shanghai. Shanghai was a 'treaty port', where Chinese laws and tariffs did not apply and where Chinese people were quite deliberately and systematically discriminated against by the (Western) civil authorities.* The city had been built on a swamp by foreign businesses and many of the buildings would not have looked out of place in Europe or the US. The 14,000 or so 'Shanghailanders' were hard-nosed merchants out to make money. Few of them had many scruples about how they did so, and their control of the city council and commercial courts meant that when it came to policing and the rule of law, Shanghai was Dodge City writ large. None of this would have appealed to Will, but it did mean that some aspects of day-to-day life were broadly familiar. This was emphatically not the case in Peking, a Chinese city through and through.

Politically, the Imperial court viewed the presence of foreigners so close to the Forbidden City, the seat of the Dragon Throne, with a profound lack of enthusiasm. Conversely, foreign governments' insistence that they should be allowed to open legations in Peking was motivated by the hope that through increased contact with decision-makers they would be able to influence the policy of the Chinese government. It was easy for the Ch'ing government to frustrate this aspiration, simply by having as little as possible to do with the diplomatic staff of the legations. The corollary was that diplomatic staff in Peking – a majority of the foreign presence, since commercial activity by foreigners was forbidden † – had little to do except, as one Chinese observer put it, "*drink cocktails and sleep with each other's wives.*"[56]

Such tales rarely lose in the telling, but it does appear from their diaries that some members of the expat community in Peking paid scant heed to the Seventh Commandment. Certainly, it was quite usual for foreign men – four out of five of the expat community were male – to 'keep' one or more Chinese mistresses, behaviour educated Chinese people bitterly resented. The other principal leisure activities of the foreigners in Peking were taking part in the races between Mongolian ponies held at the Pao Ma Chang racecourse, six miles west of Peking,‡ polo and gossip.

Most of the foreign residents of Peking viewed the Chinese in stereotypical terms, as devious, potentially dangerous, intellectually inferior and decadent descendants of the ancient civilisation whose bronzes and porcelain many of the expatriates eagerly collected. Even worse, the foreigners very often saw themselves as "*superior beings – more as lords of the country than as strangers in a strange land*"[57]. Diaries and travellers' accounts from this time regularly describe racist assaults by foreign men, some of whom, for example, would

---

* It is often said that at this time the gates of the public park in central Shanghai bore the notice "*No dogs or Chinese*". There never was such a notice, but it would have been a fair summary of the rules governing access to the park.
† For that reason, the Wagon-lits hotel, whose senior staff were mostly Swiss-German or German, was situated on the edge of the legation quarter. A number of shops selling luxury Western goods were also tolerated.
‡ The racecourse is now Lianchuachi Park in Fengtai District, Beijing.

CHAPTER 8

not hesitate to thrash a 'coolie' with a cane in the street merely for not getting out of the way fast enough.[58] Not all foreigners behaved like this, and some worked hard to understand Chinese languages and culture, but only a very few managed to bridge the yawning cultural divide between the two communities to form genuine friendships with members of the Peking gentry. In time, Will would be amongst that number, but for now he simply explored the city, soaking up local knowledge.

There was little scope for botanising in Peking, but in mid-April Will sent three boxes of anemone roots to Harry Veitch, saying that he was not sure whether this particular form would be new to Britain, but it was very beautiful and he thought it a potentially valuable plant. To Sargent, he wrote about two handsome poplars which he had come across in Peking, one with light foliage, one with dark, and promised grafts or young plants later in the year. He also mentioned willows, chestnuts and *Prunus* he had seen in Huailai and whose seed he hoped to collect in the autumn

Will's luggage reached Peking on 27 April, and he received his passport three days later. With Col. Anderson's help, he had hired mules and a cart and recruited two skilled muleteers.* The party set out for Jehol on 5 May. The hills north of the city, the Wutai-shan, had for centuries been an Imperial hunting-ground to which emperors regularly travelled from Peking, but the roadway was appalling, to the point where the cart overturned several times in the huge potholes. Although the country through which he passed was almost entirely stripped of trees, Will was glad to get away from the filth and dust of Peking and he wrote to Sargent about three new poplars seen on the journey, one with a lovely pure white stem and glossy bronze young leaves, also a sweet-scented ash and several oaks and conifers.[59] He reached Jehol, just north of the Great Wall, on 11 May, having taken six days to cover 120 miles.

A week later, Will wrote to Sargent[60] confirming reports that the old Imperial hunting-forest north of Jehol, which had until recently covered some 500 to 700 square miles, was being systematically felled and the land converted to farm-holdings† and that *"very soon [this] will be the same parched place as Peking."*‡ But he had travelled about 150 miles further north and based himself in a small village which was *"only just started"*, and whose name he had difficulty transliterating: the best he could do was *"Chin-tyu-shan"*. The village was situated at the foot of a range of hills which still contained a number of different tree species – he mentioned pines, spruce, birch, oak and two kinds of lime – and some rhododendrons and herbaceous plants, although the weather was still cold

---

* One sat astride each cart-shaft, hence the need for two men.
† This destruction of the Imperial hunting-forest illustrates the loss of control by Peking over the provinces, even those nearest to the capital. A generation earlier, the peasants would not have dared take an axe to the trees, and if they had then heads would, quite literally, have rolled.
‡ Will took it for granted that Sargent, an expert on trees and forestry, was familiar with the theory of dessication. Both men believed that clear-felling the forest would rapidly lead to a reduction in rainfall in the region.

On the Luan river, south of Jehol, May 1909.   © British Library Board

The Great Wall from the Luan river, May 1909.   © British Library Board

CHAPTER 8

Camp in the northern Wutai-shan, June 1909. The man seated in front of the tent is the 'Mafu', the head groom and foreman of Will's team of Chinese helpers. © British Library Board

and most of the latter not yet fully in leaf. He had collected herbarium specimens of what proved to be a new form of the 'bird cherry', a small tree with copious racemes of white flowers and very attractive autumn foliage and berries.* Wild peonies were also present, and said by the locals to be very vigorous (Will knew that Sargent hoped to cross wild peonies with cultivars to produce hybrids able to tolerate New England winters). He was marking trees to which he would return to gather seed and also young plants to lift in the autumn and was going to explore north in search of more interesting plants.

It appears that Will had to postpone the trip northwards due to ill-health. When he wrote to Harry Veitch on 5 June[61] he told him that he had had *"rather a bad time"* because of two abscesses on his neck, one of which had swelled to the size of an orange, but after getting them lanced he was feeling a little better. He attributed *"the brutes"* to *"treacherous water"* and was now meticulously boiling all his drinking-water.

More cheerfully, Will told Veitch that he had found and marked† a very fine cream anemone and was also sending seeds of a blue anemone, which he thought

---
* Alfred Rehder named it *P. padus var. pubescens* forma *Purdomii*: distinguishable from other varieties by the tomentose (hairy) lower surface of the leaves. It makes a spectacular spring display in the Arnold Arboretum.
† That is, marked the spot so as to be able to find the plant again in the autumn, when it had set seed.

Will's carts and mules on the road back to Peking from Cal-ceen-Wong, Mongolia, August 1909.
Arnold Arboretum Archives

"*a most saleable plant*". He was about to embark on a 'circuit' of over 200 miles around eastern and northern Weichang* on completion of which he would have completed his 'survey' of the province. Two weeks later, he wrote to Sargent[62] reporting the different trees he had encountered, but recording sadly that those trees which he had seen were *"just a few strays [...] I understand that ten years ago the area was well wooded"*. He also recorded seeing *Deutzia, Daphne* and a species of *Caragana*.

Will's next trip was to Liangpu-fu in the eastern part of the Wutai-shan, thence north into Cal-ceen-Wong, the territory of a Mongol prince, where he saw mature specimens of the famous Jehol elms, some of them 30 feet in circumference. Until fairly recently, Jehol had been famous for producing high-quality elm tables, but outside Cal-ceen-Wong *"the ground is taken up with rice and there is not a tree."*[63] He sent Sargent seeds of different elms and *Caragana*, and reported that he found the locals willing to help him gather seeds, also that the schoolmaster in Cal-ceen-Wong had promised to collect seeds for him in the autumn.

By 25 August, Will was back in Peking, and had applied for a passport to go to Wutai-shan (he mentioned to Sargent that this meant that he had had to surrender his passport for Jehol). He had returned from Weichang with a map of

---
* Weichang is the county surrounding the city of Jehol (now Chengde).

CHAPTER 8

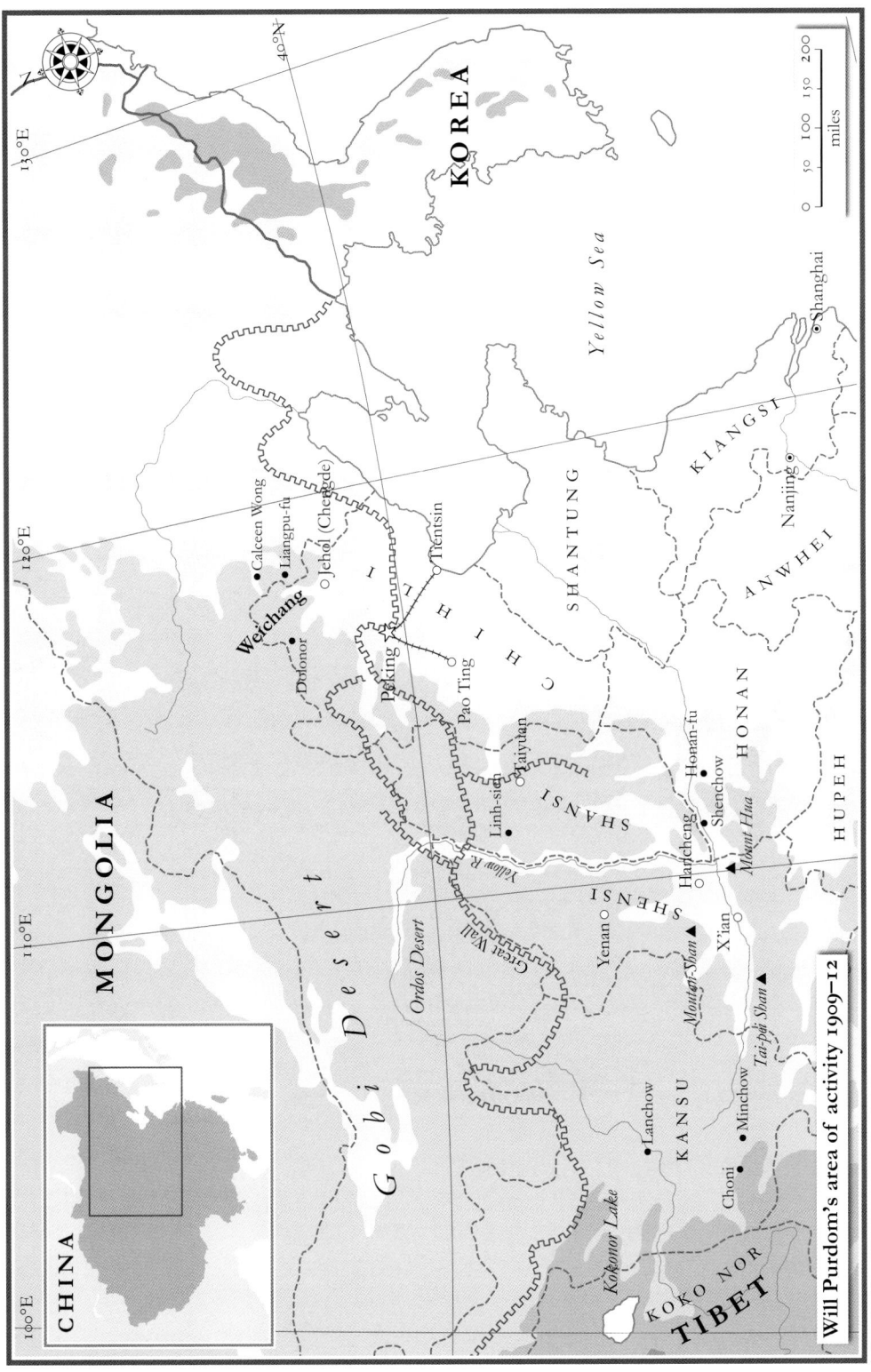

Will Purdom's area of activity 1909–12

the area, a gift from a Chinese man who was a long-term resident, to which he had added his own observations and measurements, which *"the people at the Legation thought was very good."*[64] He promised a statement of his expenses very shortly, and sent it on 4 September. The following day, he wrote a querulous letter to Harry Veitch[65] complaining that the Sanderson camera was heavy and awkward to carry. It was also very hard to stop the glass plates being smashed on the rough roads:* he would certainly recommend any traveller to bring a camera which used rolls of film, which could now be developed in Peking.†

Will received his passport for Wutai-shan in early September and left Peking by train to Pao-ting-fu, Shansi (now Baoding, Hebei Province) on the 7th. The road was too rough for carts and he hired mules for himself and his luggage. He reached

Will with priests and junior monks at a temple on the Wutai-shan, September 1909.

© British Library Board

---

* Photographic plates were supplied in custom-made wooden boxes, each plate in a sealed cardboard envelope in a separate wooden compartment. This made it almost impossible to break a plate, but the boxes, which were about the size of two shoe-boxes end-to-end, were heavy and an awkward shape. Will must have re-packed the plates to make it easier to fit them into the mule-boxes. After the 1910 season he appears to have used the maker's boxes, which he also used to send exposed plates back to London, with very few breakages in transit.

† Will persevered with the plate camera which, in the hands of a skilled user – and Will became very skilful – took much better photographs than the pocket Kodak he also carried.

the summit of the sacred mountain six days later. The mountain was stripped of vegetation – *"not a blade of grass for miles"* – and it was only in the grounds of the 20 or so temples that there was any plant life at all. The priests were pleased that Will made no difficulty about paying the customary pilgrim's levy, which entitled him and his party to lodge in one of the temples, and they allowed him to collect seed from the trees growing in the graveyards and temple grounds.[66]

It had been a very wet summer, causing Will to fear that the seeds might not be fertile, so he also gathered seedlings of the conifers and rhododendrons, as well as a small quantity of primula seeds. He returned to Peking on 6 October and immediately applied for his passport to return to Jehol to collect seeds around Weichang.

By the time the passport had been issued, autumn was drawing into winter, and north and east of Jehol Will had to contend with average temperatures just below freezing. Characteristically, he made no complaint about his own comfort, or lack of it, but was concerned that the plants in his luggage might suffer from the cold. Heavy snow caused him to divert westwards to Dolonor (now Dolon Nor) in Mongolia from where the road to Weichang was relatively clear. Dolonor, he told Harry Veitch, was *"a filthy place, and the water is not fit for pigs to drink."*[67] Will spent three days there due to sickness, and was glad to return to Jehol, and from there to Peking, where he arrived on 10 January.

Will also told Veitch that wild animals – wolf, leopard, jackal, badger, deer and wild cat – were in evidence north of Jehol and that the mountains held large numbers of highwaymen but that

"only once did I find difficulty [...] it was a night attack on a store where I slept, it was not successful and nothing further happened. Another day in the Western district a few men tried to waylay my boy [his servant] *and me in a pass but it was not a success on their part"*.

Will's reticence about these incidents – which he did not even mention to Sargent but had they involved, for example, Ernest Wilson, would surely have been exploited as an opportunity for self-aggrandisement – is typical of him and of course makes it impossible to discover anything more about the events concerned. What is clear is that Will understated the danger to which he and his party were exposed, and he may have had to use, or at least display, the lever-action rifle or the shotgun which he carried. Whilst reticence was fundamental to Will's nature, it's hard to suppress the thought that rather less understatement on his part might have helped Veitch and Sargent understand the dangers and difficulties inherent in the work he was undertaking on their behalf.

It is also noteworthy that Will said not a word to Sargent or Veitch about his lodgings during the time he was on the road searching for plants. Although Colonel Anderson had helped him procure a good strong tent, which had proved its worth in Weichang during the spring and summer,[68] for the last three months of 1909 he had been mostly living in cold, dirty and crowded inns. Again, it is hard to avoid the thought that if Veitch and Sargent had been told in greater detail about Will's accommodation during most of the 1909 collecting season, they might have been rather more supportive of their plant-hunter in China, for example by sending him a message wishing him a happy Christmas and a successful New Year.

Now he was back in Peking for the winter, Will was able to take stock of his first season's collecting. He was broadly content with what he had achieved so far. He had sent 30 parcels of seeds and plants to Veitch & Co in Chelsea* and a slightly smaller number to Boston, comprising specimens from just over 300 different plants. What Will did not understand, however, was that whilst Harry Veitch was satisfied that he had sent a reasonable quantity of seeds and plants and had made a good start to his Chinese expedition, Sargent was much less happy.

Charles Sprague Sargent was a remarkable man who created one of the world's great arboretums and herbariums in defiance of almost every possible difficulty, including a chronic shortage of funds and the near-impossibility of acquiring a large enough site on the edge of a rapidly growing and prosperous city. To describe Sargent as strong-minded is a considerable understatement and the Arnold Arboretum is a monument to what he achieved by sheer force of will. But his unfailing and immutable faith in his own judgement made him very reluctant to change his opinion, once formed (a process which rarely took any great length of time), and he was constitutionally incapable of condoning failure.

There had been a strong element of wishful thinking in Sargent's instructions regarding the first year's collecting: he had, for example, told Will that *"you will perhaps find at Wei-Chang extensive remains of the northern forests, and [...] it is not improbable that you will find many interesting and possibly entirely new plants."*[69] The background to these instructions was that in 1907 Sargent had been angry with Meyer for not sending seeds of the conifers which he had found on Wutai-shan but had failed to recognise as new species. Conveniently overlooking the fact that Meyer was first and foremost a specialist in food crops and fruit trees, Sargent had interpreted this failure of identification and the non-delivery of seeds as manifest incompetence on Meyer's part, which in turn meant that Meyer's statement that the north of China was *"scraped bare"* was not to be believed. Now Will was confirming Meyer's statement that there was nothing new to be found in north-eastern China. This was undoubtedly true, but it was not what Sargent wanted to hear.

Also, Will had not gathered seed from the new pines and larches that Sargent knew from Meyer's herbarium specimens grew on Wutai-shan and had specifically requested. True, due to a wet summer, the trees had not set seed, a problem Wilson had also encountered in 1907 and 1908,[70] and Sargent was somewhat mollified by the live specimens Will had sent, and by the substantial consignments of plants,

---

* Between 9 September 1909 and 20 April 1910, Veitch & Co received 30 parcels from Will, each containing between one and seven different packets envelopes, boxes or tins. The transit time was six to eight weeks, so Will must have put the last of these parcels in the post around late February/early March.

CHAPTER 8

cuttings and seeds Will sent from Peking to Boston. But in early May Sargent wrote to Will to say that the boxes he had received were *"generally in poor condition"* because the moss in which they were packed had been too damp and the bundles of cuttings had been tied together too tightly.* He was also critical of the size and labelling of the herbarium specimens sent by Will. Still, he concluded his letter by telling Will that *"A great many of your seeds have germinated well [...] and I think you have reason to be satisfied with your first year's work."*

Neither Sargent nor Harry Veitch had much to say about Will's photographs, but he had by now thoroughly mastered the Sanderson camera, and although late in the season the shutter had failed, he had taken a good selection of photographs of plants, landscapes and people.

Tibetan peasant girl in the Peling mountain range, 1911, clearly quite happy to be photographed by Will.   Royal Botanic Garden Edinburgh

A striking feature of Will's photos of people is the high proportion of his subjects who are laughing or smiling. Clearly, from behind the camera, Will was engaging them in good-humoured conversation.† No one, even those not actually laughing out loud, appears at all reluctant to be photographed. The contrast between Will's pictures and the sullen faces of some of the persons photographed by other travellers in China around this time, who all too often might just as well be holding placards stating that they have been ordered to stand still to have their picture taken, says more than could any words about Will's interaction with Chinese people.

On 14 January 1910, Will called on Dr Ernest Morrison, the Peking correspondent of the London *Times*, who had visited Shansi and Shensi (now Shanxi and Shaanxi) provinces the previous October. This was the area in which

---

* It cannot have helped that plant material from overseas took much longer to clear US Customs than similar material processed by the British authorities. Sargent does not give the date on which he received the boxes, but his letter concerning them is dated 3 May 1910, whereas Veitch wrote on 24 February to say that his plants – despatched from Peking at the same time as those sent to Boston – *"seemed to be in good condition."*

† It's likely that many of Will's rural subjects, who might well never have seen a foreigner before, found the mere fact of his being able to speak to them in Chinese amusing.

Sargent had instructed Will to botanise during his second collecting season, and Will wanted Morrison's advice about travelling and working there.

Will reported to Sargent that Morrison recommended he should base himself in the ancient city of Xi'an for the season, rather than, as Sargent had suggested, in the small town of Pu-Chou (now Pucheng), around 60 miles north of Xi'an. He gave Will detailed advice about the best routes to take in order to cover as much botanically unexplored territory as possible. Morrison also told Will that the Chinese officials with whom Will had come into contact during his peregrinations had reported favourably on him to the Peking authorities, which augured well for his request for a new passport for the next season's collecting.

Morrison recorded in his diary what Will had told him about the state of the northern forests:

"William Purdom, collector for James Veitch and Sons and the Arnold Arboretum, U.S.A. called and gave me a graphic account of the devastation of the splendid larch woods in the Hunting Reserve – the Wei Chang. No care is being taken & there is no conservation: a whole tree of valuable wood will be sacrificed to yield one beam of the size of a railway sleeper. Trees are felled and the stumps 3 feet high allowed to stand to collect fungus which spreads to the young neighbouring trees. Sheer ignorance combined with the reckless waste inseparable from the present Government system where public office is never regarded as public trust are accountable for the destruction. There is no system of thinning and a valuable asset is disappearing in smoke, for the people after all must keep warm and the cheapest fuel they can get is the nearest timber. Employment of an expert entrusted with adequate authority would cost the Government a few hundreds a year and would save hundreds of thousands of pounds.

There is a fine profusion of timber of many kinds but all is going to waste and he 'does feel so sorry'." [71]

In the light of subsequent events, Morrison's comment concerning the benefits that would flow from the employment by the Chinese government of expert foresters is prophetic; indeed, this discussion suggests that Will may already have been thinking about how he might secure a position which would enable him to preserve and protect the forests of China.

Will wrote to Harry Veitch asking for a few items from London, including a new shutter for his camera and some more glass plates, and arranged – not without difficulty – cash and letters of credit for the 1910 season.[*] Meanwhile, Colonel Anderson had procured for Will a copy of William Rockhill's book, *Land of the Lamas*,[†] which contained a detailed description of part of Will's proposed route

---

[*] China did not have a national currency and the principal form of money was small boat-shaped silver ingots. The standard weight of these ingots varied between provinces, and many provinces had their own units of weight, so what was an ounce of silver in Peking might be counted as either more or less elsewhere. The proof (the proportion of pure silver) required also varied. Copper coins ('cash') were used for small transactions, but were very heavy to carry.

[†] Rockhill was by now US Ambassador at St Petersburg.

CHAPTER 8

to and through Kansu Province, and had organised the making of the mule-boxes Will would need in the spring. Will also told Harry Veitch that Anderson "*was good enough to get me all the information he can get at (secret service)*" about the country Will would be crossing.

Will was unaware that whilst he was packing and despatching the harvest of the 1909 season and making preparations to collect in Kansu in 1910, Sargent had been negotiating with Ernest Wilson to persuade him to return to China. Wilson had returned to Britain in May 1909, where he had been reunited with his wife and three-year-old daughter in Kew. He had spent the next three months recovering from what had been a tiring journey home, classifying some of his introductions at Coombe Wood and cataloguing the glass-plate photographs he had brought home in his luggage before taking up a post at the Arnold Arboretum.

Wilson did not much enjoy the taxonomic work in which he was engaged at Boston and found his salary insufficient for his needs and those of his family. But Sargent refused to increase Wilson's pay unless he returned to China to collect fresh supplies of the striking *Lilium regale*, the Regal lily.

Wilson had first found this spectacularly beautiful plant in Western Szechuan in 1903, and had sent bulbs back to Harry Veitch at Coombe Wood, where they flowered in 1905, when Kew wrongly identified them as *Lilium myriophyllum* (modern synonymn *L. sulphureum*). Sargent had much admired the plant during one of his visits to London and had tasked Wilson to collect as many bulbs as possible during his 1907–09 China expedition. Wilson, who delighted in collecting very large quantities of plant material, duly gathered several thousand bulbs* in 1908, and in March 1909 he had supervised their packing in clay before despatching them to Sargent. Unfortunately, the crates of bulbs were stowed in the ship's hold immediately next to a shipment of uncured hides and they reached Boston as a foul-smelling mush: not a single bulb survived.

Sargent wanted to gain for the Arboretum the glory of introducing *L. regale* to America and was also keen to make good the $6,000 to $7,000 loss he estimated the Arboretum had incurred as a result of the loss of Wilson's 1909 consignment of bulbs.† He also wanted to secure seeds of several species of fir that he knew only from herbarium specimens, and was convinced there were further conifers, as yet unknown outside China, to be found and sent back to Boston.

In January 1910, Sargent wrote to Harry Veitch confirming that "*I have determined to send Mr Wilson back to China to collect principally seeds* [of the trees] *which he failed to obtain during his last journey*". Sargent added that he had commissioned Wilson to collect 33,500 lily bulbs (not all of them *L. regale*) and did Veitch want Wilson to collect any bulbs or seeds for him? In a manuscript PS

---

* Such behaviour would be completely unacceptable today but was standard practice at the time.
† Sargent had entered into a contract with prominent Boston nurseryman John Farquhar, who had agreed to pay 35c plus shipping costs for each viable bulb.

Sargent added *"Would you kindly not mention at present the fact that Wilson is going out to Shensi again & oblige, CSS"*.

In March 1910 Wilson returned to England with his wife and daughter, whom he installed with his parents in Birmingham before travelling to Peking by the trans-Siberian express, arriving in late May 1910. He wasted no time in proceeding to Yichang, Hubei Province, 600 miles due west of Shanghai, where he rapidly reconstituted his team of helpers and set out for Chengtu (now Chengdu) in Szechuan Province 550 miles further west.

Sargent was now sponsoring two separate plant-hunting expeditions – Wilson's in south-western China and Will's in north-western China. Both men moved around a good deal, making it impossible to give a precise figure for the geographical separation between them, but for most of the time the two parties were less than 500 miles apart. The difference between the flora of their hunting-grounds is, however, much greater than this relatively short separation might suggest. The climate of Szechuan and Hunan Provinces is sub-tropical, shading into tropical, and the annual monsoon delivers plentiful rainfall. Gansu, Shanxi, and Shaanxi Provinces share a temperate climate, shading into cold as altitude increases, with bitterly cold winters and little rainfall. Unsurprisingly, the flora of Gansu and its immediate neighbours is noticeably sparser than the sub-tropical vegetation of Szechuan.

Furthermore, the Hengduan mountains in western Szechuan are far enough south that during the last Ice Age they escaped being scraped bare by glaciers. The substantial variation in altitude created a range of different habitats, from the Tao river valley to alpine meadows and peaks, and a huge range of plants flourished there whilst those further north were wiped out by the ice. In consequence, the whole Hengduan massif is a 'biodiversity hotspot', a veritable plantsman's paradise in which a 1997 expedition by botanists from the Arnold Arboretum counted over 8,500 species of plants, spread over 1,467 genera, and 15 per cent of them endemic (found only in that confined geographical area).[*] They include over one in four of the world's species of rhododendrons (224 species), primulas (113 species), and sorbus (36 species), 41 species of cotoneaster, and more than 30 species from the rose family – the list goes on and on.[72]

In contrast, plant biodiversity in what is now called the Qinling mountain range in Gansu province has been measured at 1,044 species of plants, and in south-east Gansu at 2,458 species, of which just 29 are endemic.[73]

Other surveys may produce marginally greater or lesser figures than the above, but what is clear is that Wilson was plant-hunting in an area containing somewhere between three and a half and eight times the number of plant species to be found in the area to which Will had been sent by Sargent and Veitch. In light

---

[*] By way of comparison, the British Isles presently have 1,443 species from 308 genera, only 1.2 per cent of them endemic.

CHAPTER 8

An inn-keeper and his staff at a hostelry on the way to Shensi in April 1910. As the clockwork self-timer clicks, the woman on the left in the back row is telling the young girl on the extreme left not to peep out! All three men in the front row are visibly amused by the dialogue going on behind them.

© British Library Board

Will and the Mafu passing through a gate in the Great Wall, Shensi, April/May 1910.

© British Library Board

of this, it would have been surprising if Wilson had not collected and sent back to Boston specimens and seeds of more species than Will.

To be fair to Sargent, the detailed data about biodiversity in the Hengduan range would not be available for another half-century. But Wilson had made no secret of the remarkable wealth of plant life in south-western China, for example telling the Kew Guild in 1901 that the area was *"a botanical paradise"* and subsequently describing the variety of flowering trees and shrubs of Western Hupeh as *"well-nigh infinite"*. The different climates of Wilson's collecting-area and Will's should also have given Sargent a strong hint that there were likely to be fewer species of plants growing in the latter, so that, other things being equal, Wilson would send back more, and more diverse, plant material than Will. Sargent's refusal to accept this is characteristic of his ferocious self-confidence but does not reflect well on his personal judgement.

Chapter 9
The 1910 season

On 15 April 1910, Will and his staff left Peking by train for Taiyuan, the capital of Shansi (now Shanxi) province in north China. From there, Will decided that the first staging-post on the way to Shensi (now Shaanxi) province in the north-west of China would be the ancient city of Xi'an. As Ernest Morrison had suggested, rather than follow the main road between Taiyuan and Xi'an, Will and his team struck off on foot north and west over the mountains to Linh-sien (now Linxian), then a little further west to cross the Yellow River near modern Jiaxien before turning south to Yenan (now Yan'an) and then south along the west bank of the Yellow River to Hangcheng, where Sargent's written instructions to Will asserted that "there is a hill called Moutan-shan on which grows the wild Moutan or Tree Peony (see Bretschneider)."*

This last claim was, even by Sargent's standards, an extreme instance of wishful thinking. What Bretschneider's magisterial 1898 work *European Botanical Discoveries in China* actually says about Moutan-shan is that

"a Chinese description of Shensi Province published about 200 years ago [i.e. in the late 17th century] reports that [...] there is a hill called Moutan-shan† *where the moutan tree grows in great profusion, in a wild state. In spring, when these trees are in blossom, the whole hill appears tinged with red, and the air round about for a distance of ten li* [five miles] *is filled with fragrance."*

It is entirely understandable that Sargent should have found the description of Moutan-shan alluring and that he should have asked Will to investigate the site, but his determined belief that nothing had changed there in the last two hundred-odd years was typically Panglossian in its optimism.

In any event, Will's route, roughly the two sides of a triangle of which the direct route from Taiyuan to Xi'an constituted the third side, meant that the party had to make their way over 500 miles across largely trackless and thickly wooded terrain in which the complete absence of any other accommodation meant that they had to camp. Tough going though it was, the route had the great merit of being through botanically unexplored country.

Will reached Xi'an on 7 June, and reported to Veitch and Sargent that although in places he had found it difficult to force a passage through the woods, the district

---
* Emil Bretschneider (1833–1901) was a Russian physician and sinologist of German ethnicity. He was the doctor at the Russian Legation in Peking from 1880 to 1883, during which time he sent dried herbarium specimens to Kew. He published, in English, several works on Chinese history, of which the best-known is his two-volume account of the botanical exploration of China by Westerners.
† 'Moutan' (in pinyin 'Mudan') is Chinese for tree peony, and 'shan' means mountain.

CHAPTER 9

Will's camp on or near the Moutan-shan, May 1910. His rifle is resting against a press for preparing herbarium specimens.   © British Library Board

north of Hancheng had been very pretty, with wild pear in flower, also a *Malus* with pear-like fruit, two kinds of *Syringa* (lilac), and a poplar with attractive bronze foliage. He also mentioned abelias, lacebark pines (*Pinus bungeana*), viburnum, sea buckthorn and rhododendrons. He had gone as far west as the edge of the Ordos desert, where he had been impressed by the spectacle of the sand banked up against the Great Wall, before turning south-east through "*a blank piece of country not on the map*".

Local officials had tried to make him take the main road to Xi'an but he "*managed to dodge them*". He spent five days making his way through the woods, which were alive with game, including wild boar, deer, wolf and leopard: the presence of the latter meant that they had to stand watches at night to keep the mules safe. When he reached the Moutan-shan, there was not a single peony to be found,[*] and locals had told him that none had grown there within living memory, but he intended to return at the end of the season to collect the seeds of other plants, and he would search further west of the hill. He had in any case already found two other sites, to the west and to the south of Yenan, where peonies were growing wild.

---

[*] Will took photographs of the bare hillside so as to leave Sargent in no possible doubt about the absence of peonies, or much else, growing there.

Will sent both Veitch and Sargent a box of seeds (obviously, although recently gathered, these were last season's) and specimens, including a few seeds of a tree he commended to both of them as having light blue flowers with dark purple stamens, *"very showy, gave off a nice scent, and was in flower at the end of May."** He asked if Veitch could have a map copied for him *"as most of my route is on new ground and it could be filled in if one had a large copy [map] of Shensi"*.

By happy coincidence, on 8 June, the day after Will had reached Xi'an, Sir Alexander Hosie arrived there, and lodged at the inn adjacent to the one where Will was staying. The previous year Sir Alexander had represented Britain at the negotiation of an agreement between the Chinese and British governments that the export of opium from Bengal to China would be progressively reduced and from 1917 completely banned. In exchange, the (British) India Office required China to reduce the cultivation of opium poppy in China, also culminating in a complete ban. Sir Alexander was now visiting the chief opium-growing regions of China to monitor progress. He and Will were happy to renew their acquaintance, and Will was able to tell Sir Alexander that poppy was being widely and openly cultivated in the district north-west of Hancheng. Hosie asked Will in due course to write with his observations about the extent to which poppy was being grown south and west of Xi'an, an area through which Will would pass but which Sir Alexander did not have time to visit.†

Will and the Mafu, summer 1910. Note the revolver holster just in front of Will's right hand.
Royal Botanic Garden Edinburgh

Hosie left Xi'an on 14 June and the following day Will's mail, including Sargent's letter of 3 May complaining about the packing of cuttings and Will's herbarium specimens, caught up with him. Will wrote an emollient letter to Sargent promising that in future the plants would be packed drier. He also referred to an earlier letter from Weichang (which Sargent does not appear to have received) saying that

---

* Possibly *Paulownia glabrata*: *Plantae Wilsonianae* records that this tree was identified by Will in 1910. Modern synonymn *P. tomentosa* var. *tsinglingsensis* (Pai) Gong Tong.

† Twelve days later, Will wrote to Sir Alexander informing him that once he got ten miles away from X'ian he found poppies growing on around one-third of the available agricultural land. He noted that *"nearly every man smokes opium and, worse than that, quite a large number of the womenfolk are addicted"*.

CHAPTER 9

Deforestation in Shansi Province. © British Library Board

the bulk of the herbarium specimens would follow at the end of 1910, with the collection from Shansi (modern Shaanxi) Province – the dried specimens Sargent was unhappy about had been intended only to give him an idea of what the plants looked like. He sent Sargent a photograph of the blue-flowered tree of which he had posted seeds a few days previously and told him about the *"very interesting"* wild pear trees in Shansi Province, which Sargent had specifically tasked Will to collect. He hoped to collect seeds and cuttings later in the year, when he went back to Yenan-fu for moutan seeds. He also asked Sargent to confirm receipt of the various conifers he had sent earlier in the year and for a list of which conifers, poplars and elms had germinated. He needed this information before he wrote to a man in Weichang who had promised to collect more seed. Will closed by reporting that he had *"fixed up mules"* and was off to the Tsingling (now Qinling) mountain range the following day, 16 June.

Even now, when access is far easier than it was a century ago, botanising in the Qinling mountains is not for the faint-hearted. The range runs east–west for nearly a thousand miles and is between 4,000 feet and 12,500 feet high. The mountains act as a barrier to the annual monsoon winds blowing northwards from India and are also the watershed between the Yellow River basin of northern China, which receives little rain and suffers cold winters, and the Yangtze river basin of southern

The summit of the Qinling mountain range, above the tree-line. © British Library Board

China, which is much wetter and enjoys warm winters. In Will's time, the southern slopes were densely wooded, with deciduous trees giving way to evergreens above 4,000 feet. The terrain is fissured with narrow, deep ravines, and the altitude causes shortness of breath, sometimes full-blown altitude sickness.

Will took four days to reach the foothills of the Tai-Pei-Shan (now Taibai Shan), at which point the mules could go no further. Continuous heavy rain swelled mountain streams into raging torrents and he had difficulty in recruiting porters to carry baggage, but after a week he and his party had established camp more than 6,000 feet up the mountain and were busy collecting seeds and specimens. July was the season for pilgrims to visit the temples in the mountains and the priests there spread the word that the bad weather was due to Will's impious presence. As the pilgrims became increasingly hostile, Will wrote to Harry Veitch that *"things began to get rather serious"*, and he sent a man down the mountain to Xi'an with a message to transmit to Colonel Anderson in Peking. Anderson promptly complained to the Chinese Foreign Ministry, who ordered the authorities at Xi'an to inform the priests that Will was under the protection of the government. This immediately brought about a change of attitude; Will told Harry Veitch, *"each man who was hostile before could be seen carrying a few nuts to my tent, or a few flowers"*.

CHAPTER 9

There is no reason to doubt the accuracy of Will's account of this exchange with Anderson, but it is not the whole story. At the end of the year, Anderson wrote to his friend and former colleague at the Peking Legation, Jack Garnett, who had been posted by the Foreign Office to St Petersburg,[74] mentioning that in September he had been *"away in Shensi on work"*, and he appears not to have returned to Peking until mid-December. Anderson's movements around Shensi province and what he was doing there are unknown. But his later comment in a letter to a friend in London[75] that *"Mr Purdom is an excellent fellow and a most loyal one"** suggests that Will supported Anderson in whatever *sub rosa* activities he was engaged in.

Will and his team collected for several weeks on the southern slopes of the Tai-Pei-Shan massif, and climbed to the summit itself, where he noted that although nothing grew save very small alpines, the colours made *"a wonderful effect"*. He told Harry Veitch that he had *"made quite a collection of stuff, some of which to my mind will be very serviceable"*.

One of the plants Will collected was a new rhododendron whose dark pink buds turned into beautiful pure white flowers: when Will's herbarium specimen reached Boston, Alfred Rehder the Arnold Arboretum taxonomist, identified it as a previously unknown species and named it *Rhododendron purdomii*.[†]

Unfortunately, towards the end of August Will's interpreter[‡] and two other men fell ill and Will was obliged to take them back to Xi'an for medical treatment before rushing back to the Tai-Pei-Shan. There he gathered a large quantity of seeds before going on to Yenan, a difficult journey because the roads were flooded.

At Yenan, Will found that the two local Chinese men whom he had asked to collect seed for him had gathered 530 moutan seeds, around two pounds of wild pear-tree seeds, and seeds from various other local plants. He arranged for them to gather moutan seeds next year and to forward them via a friendly missionary. On 30 October Will wrote telling Sargent that he had found eight moutan plants growing wild on the Tai-Pei-Shan, but unfortunately they had not set seed, probably due to the wet summer. Even after a second visit, he had found no peonies growing on the Moutan-shan, but he hoped that the fact of having found one in three separate locations in Shensi meant that this was the true wild peony.[§] He drew Sargent's

---

* Meaning that Will was loyal to Britain and the Empire, *i.e.* a patriotic Englishman.
† The taxonomic classification of this rhododendron is controversial and is considered in detail by Ejder, Ridderlöf and Salomonsson in the article cited in the bibliography.
‡ The dialect spoken in Shansi Province was a quite different language from that spoken in Peking and the interpreter would have acted as translator for Will's team of assistants as well as for Will himself. A recurring theme in the narratives of foreign visitors to this and other remote northern and western Chinese provinces is the difficulty they experienced in finding persons who could speak the local tongue and a European language and who were willing to travel to areas considered by most educated Chinese people to be uncouth and potentially dangerous.
§ Tree peonies have been grown in Chinese gardens for centuries, and there are hundreds of ancient cultivars bred by gardeners in China and Japan. Inevitably, some of these have established themselves in the wild. The quest for the original species from which all these cultivars ultimately derive – one or more of which or may not have survived in the wild – has long preoccupied both Chinese and Western botanists. Sargent's keen personal interest in obtaining the wild form was driven less by a thirst for botanical knowledge than by the hope that he could use it to produce peony cultivars that would be fully hardy in Boston.

attention to a further species of pear which he thought was a good form* and asked him to identify samples from a plant that Kew said was *Crataegus pyrathanca*,† an identification he thought was incorrect. Will closed his letter by saying that he would now explore the area south and east of Yenan and west of the Yellow River before returning to Peking a little before Christmas, when he would ship the seeds he had gathered to Sargent and Veitch. He concluded by saying that he hoped the plants he had collected that season would "*prove valuable to science and agriculture*".

It is likely that Sargent received approximately simultaneously the letter recording what he regarded as Will's statement of heavily qualified success in 1910 and a letter from Ernest Wilson reporting that he had succeeded in marking the location of thousands of *Lilium regale* bulbs,‡ and that most of the conifers of which Sargent wanted seeds had been found and they appeared to be setting seeds this year, but that on 3 September Wilson had been caught up in a landslide west of Chengdu and had sustained serious injuries to his right leg. Fortunately, Wilson said, his detailed plans for the rest of the year were well understood by his staff so "*the accident cannot involve the expedition in entire failure*".

To say that Sargent and Wilson were personal friends would be an exaggeration, but their relationship was based on mutual respect and was as cordial as any working relationship between Sargent and another man. Sargent was sincerely distressed by Wilson's news. He vented his feelings on Will, to whom he replied by return of post. After tersely confirming that Will was right and Kew wrong about the putative *Crataegus*, which was in fact *Plagiospermum sinense*, Sargent expressed disappointment that Will had not sent at least some seeds "*as fast as you gathered them*", since if they were held back until Christmas they would lose vitality. He also expressed surprise and regret that Will had returned to Peking rather than "*pushing westward so as to have reached Kansu in time to begin work in the very early spring*". He complained that he and Harry Veitch had received practically nothing of the year's collecting and were "*entirely in the dark*" as to Will's success (or, by implication, his lack of success) and urged him to leave Peking at once "*in the hope of reaching Kansu in time, if possible, for good spring work*". He hoped Will would write more frequently and at greater length: he did not like the state of ignorance in which Will's short letters left him and Harry Veitch. Finally, Sargent complained that Will had not sent seeds of *Aesculus chinensis*§ "*which is cultivated in some of the temple grounds in Peking and probably in other places*", nor the wild pear from Jehol: he hoped to receive both next year.

---

* That is, a species or hybrid with characteristics making it worth cultivating.

† Harry Veitch had presumably sent a sample to Kew and advised Will of the identification, though his letter doing so does not survive. Sargent was fascinated by *Crataegus*, and wrote at considerable length about the genus in his encyclopaedic work on American trees.

‡ The plants grow 18 inches high in the spring, but it is only after they have died away in the autumn and become quite invisible that the bulbs can be lifted. Wilson's helpers had to mark each one, then return four months later to dig them up.

§ British readers may know this handsome tree as the Chinese horse-chestnut, and Americans as the Chinese buckeye.

CHAPTER 9

Meanwhile, Wilson's staff carried him on a stretcher to Chengdu. The initial diagnosis was that that his leg would have to be amputated, but after less drastic surgery and a further two months of painful treatment he was sufficiently recovered to travel to Shanghai, thence back to Boston, where he arrived in March 1911.

Sargent's letter to Will took the usual six weeks to reach Peking, and crossed with another letter from Will written from Peking on 12 January, by which time Will had heard that Wilson had returned to China in 1910 and what had befallen him.

Will took his cue from Sargent and did not mention Wilson in his letter, but he struck an unusually bullish (for Will) tone. He expressed some satisfaction with the harvest of plants from Shansi. He had sent *"quite a number"* of packets of seeds that should reach Sargent shortly, and would soon post some young plants of larch and other species of which he had not been able to gather seeds. He had made a nice collection of ferns which he hoped would interest Sargent's correspondent Dr Christ.* The vegetation of the previously unexplored terrain on the Tai-Pei-Shan had proved interesting, and he had worked out a route to take him to Kansu early next year through similarly unexplored territory. He promised a list of his expenses very shortly, and asked Sargent if his second half-year's remittance for 1911 could be sent a month early, so that he could collect funds from Lanchow at the end of May.

Ten days after posting this letter, on 22 January 1911, Will received Sargent's diatribe of December 16th. Will replied the same day, in a rambling letter combining apology and self-justification. He pointed out that the two areas in which he had collected in 1910, Yenan and the Tai-Pei-Shan, were far apart. The roads had been bad due to heavy rain but he had nonetheless managed to gather seed on the Tai-Pei Shan and to travel to Yenan in time also to be able to gather interesting seeds there, after which he had proceeded straight to Xi'an to send them to Boston and London. (The sub-text is that if he had travelled straight from the mountain to Xi'an to post the first harvest of seeds, it would then have been too late to proceed to Yenan to gather seeds, but it might have been clearer to Sargent if Will had spelled this out.) His decision to return to Peking was partly due to the need for him personally to pack and despatch the seeds and plants he was sending to Sargent and Harry Veitch. The sickness of his interpreter and the two best members of this team and the hostility of the locals on the Tai-Pei-Shan had made life difficult, and frustrated his plans to split his team between Yenan and the Tai-Pei-Shan. He was sorry if the dried herbarium material was deficient, but he was not a trained botanist and was doing his best.†

As far as going direct to Kansu was concerned, Will said that his *"followers"* would not have wanted to go straight from Xi'an to Kansu, and he needed to re-supply and to replace much of his equipment, most of which had been stolen by highwaymen on

---

* Dr Hermann Christ of Basel, Switzerland (1833–1933), a noted pteridologist.

† By any reasonable definition, Will *was* a trained botanist, but his training in preparing herbarium specimens had been limited to a fortnight in Boston.

the Tsingling range. In any case, even if he had gone straight there, he would not have been able to get further than the border of Kansu. (This last was undoubtedly true, but it might have been a good idea for Will to spell out that the mountain passes were closed by snow between late November and, at the earliest, mid-March.)

Will closed by saying that plants of larch, pine, red-barked birch,* and some others, were on their way to Boston. He hoped they would arrive in good condition and regretted any inconvenience he might have caused Sargent.

Will's letter again crossed with one from Sargent, written on 15 February. Sargent was somewhat mollified by a large consignment of seeds which had arrived in Boston in good condition, including the moutan seeds, though he would have liked a larger quantity of the latter. He endorsed Will's decision to travel through unexplored territory to reach Kansu rather than taking *"the high-road"*, wished him success in the new season and confirmed that as requested he had arranged for Will's expenses to be paid a month early.

A fortnight after writing this letter, Sargent received Will's letter of 22 January and promptly sent a copy to Harry Veitch. Sargent's covering letter to Veitch does not survive, but it appears that he was unhappy with Will's latest communication and with his general conduct. Sargent's negativity towards Will may have been partly a reaction to Wilson's imminent arrival in Boston and the arrangements Sargent was making for him to be hospitalised there for further treatment.† Sargent felt guilty for having sent Wilson into harm's way and this caused him to draw comparisons which were not to Will's advantage.

Veitch replied to Sargent on 25 March and urged him to be more indulgent, reporting that a mutual friend of his and Colonel Anderson had received a letter from the Colonel in Peking praising Will, who was by now *"off on another botanical hunt"*.

Sargent did not reply directly to Veitch's letter, but he wrote to him saying that *"his [Purdom's] seedling plants came in fair condition"* and expressing satisfaction that Will had sent peony roots to Chelsea. He hoped that they would receive special care *"as I consider this the most important and most valuable of Purdom's introductions"*.

This last judgement reflected Sargent's keen desire to obtain specimens of 'the true Moutan peony', partly because he hoped it would withstand the New England winters and could be used to produce cultivars which would be equally hardy and partly because he craved the status that would accrue to the Arnold Arboretum if it could showcase specimens of the true, the original, peony.‡

---

* *Betula utilis* ssp. *albosinensis*, a strikingly beautiful tree of which in the 1911 season Will would collect a large quantity of seed.
† Wilson spent six weeks in hospital, emerging with a permanently crooked right leg an inch shorter than his left, after which he travelled to Britain for a further three months of physiotherapy. Although never known to complain about what he called his 'lily limp', he was left lame for life.
‡ Enthusiastic admirers and collectors of different plant species attach much greater value to specimens of the original wild species than to cultivars, hybrids produced by gardeners or horticulturalists.

CHAPTER 9

Will's peony roots arrived in Chelsea on 20 January 1911 under collection number 545. There is no record of how many roots the parcel contained. On 24 January some were sent to Coombe Wood and some to Langley.

Harry Veitch sold his share of the peonies (including herbaceous peonies grown from the seed Will had sent in late 1909) and which he had named *Paeonia Veitchii* (now *P. anomala* ssp. *veitchii*) to Robert Woodward, a leading peony enthusiast who grew them at his home, Warley Castle in Worcestershire. Woodward's plants grew little more than a foot tall and had large pink petals surrounding showy yellow anthers. They are currently cultivated under the name *Paeonia veitchii* var. *woodwardia*.

Some plants were sent by Veitch to Boston, where they were cultivated in Sargent's personal garden and were identified by Alfred Rehder as *Paeonia suffruticosa*.* Unfortunately there is no record whether these were grown from the seed Will had sent in late 1909, or the roots he sent in late 1910. Rehder later (1913) recorded in *Plantae Wilsonianae* that *"with the evidence of Purdom's collection there can be no doubt that P. suffruticosa is a native of north-western China and was introduced from there into western China and Japan."* One is left wondering why neither Sargent nor Harry Veitch publicised this discovery, particularly as Sargent had specifically tasked Will to collect wild peonies.†

---

* *P. suffruticosa* is not now considered a species in its own right, but rather a variant or variants of *P. veitchii*.
† It is possible that Sargent didn't believe Will's plants to be a true wild species, because like many other botanists he thought the original peony was white with a dark purple/red blush in the centre of the flower. There is a putatively wild peony matching that description and in 1924 the Arnold Arboretum funded an expedition to Kansu and Tibet by the Austrian-American collector Joseph Rock, who based himself for two years in the lamasery at Choni, Kansu Province. In 1925 or 1926 Rock sent Sargent seed from a peony he found growing in the lamasery garden. This was (wrongly) accepted as a wild species and named *P. rockii*. The convoluted and sometimes confused story of the different plants identified at various times as *P. rockii* is set out in detail in McLewin and Chen (see bibliography).

Chapter 10
1911: Things fall apart

Will left Peking for Xi'an on 20 February, having written to Sargent two days previously that he would be in touch towards the end of the collecting season and giving Sargent *"the lay of the land"* so that he and Harry Veitch could decide whether they wanted to extend Will's three-year contract, which would otherwise expire at the end of February 1912. Will had made appropriate arrangements with his Chinese staff in case Sargent wanted them to overwinter in Kansu, but a decision on this would have to be taken by the end of the year. Meanwhile, Will enclosed accounts for his expenses incurred in 1910.

A month later, Will was in Xi'an, writing to both Veitch and Sargent that the roads were still snow-bound but he was off to Kansu on 22 March. He was pleased that Sargent had received the moutan seeds safely (it is not clear how Will knew this, no relevant letter from Veitch or Sargent survives). He made some detailed comments about the plants of which he had collected seeds the previous year and told Sargent that he had made a list of the plants Przewalski[*] had collected in Kansu. He hoped to find the seeds of many of them, also seeds of the pears the Russian explorer, 'Kougloft' had found in Suite-chow (Suzhou?) in 1907.[†] He repeated that, notwithstanding efforts by Chinese officials to discourage foreigners from going there at all, he would be guided by Sargent as to the time to be spent in Kansu.

Four weeks later, on 19 April, Will wrote again to Sargent from Minchow (now Minxian) in Kansu, 300 miles west of Xi'an. The countryside immediately around the town was mostly grassland, where he had hopes of early primulas and *Meconopsis* (poppies). He had met William Christie, a local missionary and Tibetan scholar who had the previous year met two members of the expedition funded by the Duke of Bedford to collect specimens from south-west China and Tibet for the British Museum of Natural History. They were the expedition doctor, Jack Smith, and 24-year-old Frank Kingdon-Ward, on leave from his job as a schoolmaster in Shanghai. Although the focus of the Bedford expedition was firmly on mammals, Kingdon-Ward, a botanist who went on to become a distinguished plant-hunter and explorer, also made a small collection of herbarium specimens.

---

[*] Nicholas Przewalski (1839–1888) was a Russian explorer of Polish ethnicity who made four journeys into central and eastern Asia between 1870 and 1885, although he never achieved his lifelong ambition of reaching the Tibetan capital, Lhasa. His name is known in the West principally because of the eponymous Przewalski's horse, a remote ancestor of the domestic horse of which in the 1870s he brought hides and bones from Mongolia to St Petersburg. He is also commemorated in the scientific names of more than 80 plants, five lizards, and a gazelle. He is remembered in Russia for the persistent urban myth, based on facial resemblance, that he was Joseph Stalin's natural father.
[†] Will was referring to Przewalski's young disciple, Pyotr Kozlov (1863–1935).

CHAPTER 10

Christie was not especially interested in botany but on the basis of his conversations with Kingdon-Ward the previous year he advised Will that the hills to the west, part of the small semi-independent Tibetan principality of Choni (now Joni or Zhuoni), were well wooded and a good hunting-ground for new plants. Accordingly, Will intended to base himself in the district for the 1911 season. Christie (who had lived in Choni between 1904 and 1907 and continued regularly to visit the town) was friendly with the powerful hereditary ruler of Choni, which might be helpful. Will added that there was now an international post office in Minchow: Sargent could write to him there and the letters would reach him more quickly than if sent to Peking.

On the same day, Will wrote to Harry Veitch, giving him a little more detail about the weather (very cold, despite bright sunshine) and the local flora (not yet emerging, but he had hopes of poppies and primulas). He grumbled that the locals constantly tried to take financial advantage of him, but repeated the good news concerning the post office at Minchow: if Veitch would confirm safe receipt of Will's letter this would pave the way for sending seeds by mail, rather than sending them later in the season by ship, with the bulkier plant material.

Sargent received Will's letter on June 10 and replied the same day, saying that he was very pleased to hear Will had arrived safely in Kansu, congratulating him on getting there early in the season and expressing his confidence that Will would find interesting plants in what was *"practically a virgin country"*.

On 9 June, Harry Veitch wrote to Sargent saying that in his view Will's contract should cease at the end of the collecting season. He was not in any way blaming Will – *"If the trees and plants were not there, he could not send them"* – but the fact was that Will had not made any very striking addition to the Veitch stocks of plants or that of the Arboretum and this was unlikely to change. Sargent replied saying that he agreed and would shortly write to Will telling him that on his return to Peking from western China he should close up his affairs and return to England: Sargent saw no need for him for go home via the United States.

Shortly after his arrival in Minchow, Will had received a friendly letter from Ernest Morrison, regretting that he had been on leave in Britain when Will had overwintered in Peking, so that they had not been able to meet. Morrison had been thinking about what Will had told him (in 1910) about deforestation in the former Imperial hunting-grounds and thought that *"if we could induce the Chinese to take some scientific interest in their arboricultural resources, it would be a good step."* He asked Will to send him some notes of *"what you are doing and what you are finding and the conditions you are observing"* to provide him with the data he would need to write in the *Times* about *"the economic botany of China and […] tree cultivation, forest destruction and the effect of this destruction upon climatic conditions"*. Morrison closed by hoping that Will was being well treated by the local people, though he had few concerns on that score: *"you know so well how to deal with them that you are always well treated."*

Will took his time about replying to Morrison, but did so on 23 July. He apologised for the delay – *"I have been away in the wilds"* – and gave Morrison a summary account of his travels and botanising that season. He also sent him copies of an anti-foreigner poster which had been put up in Kweite, 150 miles east of Lanchow, and of the *"rather weak"* official riposte, as well as a printed pamphlet circulating in the region: these were *"causing a good deal of unrest [...] and quite an anti-foreign feeling along the border"*. Will suggested that *"a word in time"* to the authorities in Peking might save a good deal of trouble in the future. Although he himself continued to receive friendly treatment *"the people seem quite disturbed since the circulation of these silly writings."* Morrison replied thanking Will for his letter: he had passed on copies of the offending documents and *"steps will be taken to prevent any ill effects following on from these incredibly filthy placards."* He urged Will not to hesitate if he (Morrison) could do anything for him: *"you have only to ask me, and I will do it gladly."*

Yang-Chi-Ching, the young hereditary ruler of Choni, in 1911.
Royal Botanic Garden Edinburgh

Will botanised around Minchow for a further two months, collecting six different primulas, including *P. maximowiczii*, three Meconopsis (one large yellow form, one blue, and one red), as well as a yellow Potentilla. He befriended some Swedish-American missionaries,[*] who helped him get to know the local people, before moving his base to Choni, which proved to be a small town on the border between China and Tibet ruled by a 22-year-old hereditary local prince Yang Chi-Ching.[†] Yang was also the Abbot of the Changtingssu monastery next to Choni and controlled a swathe of territory to the north and west.

---

[*] Probably the Reverend Jens Rommen and his family, who were, like Christie, members of the US-based Christian and Missionary Alliance. Rommen shared Will's interest in hunting and was unusual amongst the missionary community insofar as he owned a rifle and knew how to use it.
[†] Yang, whose Tibetan name was Lobsang Tendzin Namygal Trinle Dorje, collected taxes and remitted them to Peking (or, as the Chinese court might have preferred to put it, paid an annual tribute) and in exchange was allowed a free hand in his domain.

CHAPTER 10

Relations between China, Tibet and British India and the internal situation of Tibet in the first decades of the 20th century were complex and confused. Very briefly, China had for centuries used Tibet as a buffer state to keep the Russians well away from the Chinese border. The Tibetan government (the country was a feudal state controlled by an aristocratic elite who also provided the senior clergy of the powerful Buddhist church) was autonomous as regards internal matters, but the Peking 'Ambassador' to Lhasa dictated foreign policy, which was to exclude all non-Chinese foreigners and refuse any dialogue with their governments. The British went along with this because they also wanted to use Tibet as a buffer state, between Russia and India. But when, in 1902, the Tibetans opened a dialogue with Moscow, Indian army officer Francis Younghusband led a column of troops from Sikkim in India to the holy Tibetan capital city, Lhasa. He used Maxim guns to crush all opposition and in September 1904 the Dalai Lama signed a 'treaty of friendship' agreeing that Tibet would in the future have no dealings with any foreign power without Britain's consent.

The Ch'ing government was dismayed by the British ascendancy in Tibet and responded by re-drawing the border between Tibet and China further to the west* and by deploying troops to assert Chinese sovereignty over 'Outer Tibet'. The soldiers concerned were Muslims from Kansu whose community had a long history of conflict with both the Tibetan Buddhists and the animist semi-nomadic tribes who occupied the high pastures.

What all this meant for Will was that when he crossed the Yao river into Choni he was entering a region where different ethnic and religious groups were very close to being permanently at war with one another and where the violent robbery of any strangers who did not have a sufficiently large armed escort to deter attack was considered to be completely normal. At the same time, Will's Imperial Chinese passport unequivocally required local officials and administrators, Tibetan as well as Chinese, to ensure his safety at all times, and all concerned well knew that there were more than sufficient Chinese troops in the region to make an example of any community where harm befell him. Their commander, General Ma Ch'i (pinyin Ma Qi), was ruthless in punishing anyone who defied his authority or that of Peking, so that the passport provided substantial protection. Still, Will knew he had to tread very carefully.†

Happily, Will was well-received by the prince of Choni. Yang was curious about the world outside Tibet and he welcomed the opportunity presented by Will's stay in Choni to engage him in conversation. Despite his friendship with William

---

* Hence Choni, which had always been considered to be just over the Tibetan side of the border, at a stroke found itself labelled part of Kansu province.
† Just how carefully is illustrated by what happened on Christmas Eve 1908 to the British explorer John Weston Brooke, when he was discussing with a local chief the fee he should pay to cross the latter's territory in south-eastern Tibet. They agreed a figure, and Brooke "*in a friendly way*" clapped the chief on the shoulder. This is a deadly insult amongst the Yi ethnic group, and the chief instantly slashed at Brooke with his sword. Brooke drew his revolver and shot the chief dead, but he and all but two of his party were killed.

1911: THINGS FALL APART

Christie, Yang may have found Will a more interesting interlocutor than Christie or any other staff from the mission station whose presence he rather reluctantly tolerated in the town.

Will and Yang were both committed to protecting the local forests, a constant cause of conflict between Yang and the Tebbu people who grazed their sheep and yaks in the western part of Yang's domain. Although Will cannot have approved of the despotic, often cruel, methods by which Yang enforced his authority, they

The main street, Choni town. Royal Botanic Garden Edinburgh

CHAPTER 10

Will with the escort of soldiers provided by the authorities for his protection while botanising in Kansu in the summer of 1911. Lakeland Horticultural Society

appear to have forged a reasonably cordial relationship, and Yang provided Will with an escort of soldiers for his trips outside the town.

In and around Choni, Will gathered seven more primulas and six species of shrubby honeysuckle. He told Harry Veitch that, although "*nomads, robbers and hostile Tibetans*" made it "*very difficult to get over*" the region, during the following week he would be making a trip to the heavily wooded mountains west of the town in the company of a Tibetan guide.[76]

It's likely that a picaresque Choni resident with whom Will had become friendly, Acho-che-ro, a reformed highway robber who was reputed to have killed five men before converting to Christianity and abandoning his former trade,[77] helped him identify a trustworthy companion who would act as a guide. He also made Will the gift of a traditional Tibetan sword for his personal protection.[78] This was an essential accessory for anyone visiting the Tebbu, of whom the botanist and ethnographer Joseph Rock, who overwintered in Choni in 1925/6 and 1926/7, said "*he that ventures into Tebbu land, let him beware; it is a robber's den inhabited by cruel, revengeful, suspicious people who carry long swords with hand always on the hilt when talking with a stranger or even a neighbour.*"[79]

Fortunately, Will's trip passed off peacefully, and in August he wrote to Sargent reporting that he had found several interesting plants, of which the best were a species of shrubby honeysuckle that produced edible blue-black fruit[*] (he

---

[*] *Lonicera caerulea*, now grown and marketed in the West as 'honeyberry'.

1911: THINGS FALL APART

stipulated that it was eaten both by Chinese people and foreign missionaries) and a *"very showy"* groundsel, finer than any he had previously seen. He was about to embark on a further trip to the high mountains to the north from which he would loop further west, returning to Choni through the area already visited.

Will still had hopes of visiting the virgin evergreen forest around the monastery of Labrang, but the lamas refused to allow him to travel there: he told Sargent that there was no point in appealing to Chinese officials, who had no real authority in the region.* The large area to be covered and the difficulties of travel meant that he would not be able to collect in the north of Kansu in the current season, and he awaited Veitch's decision on whether at the end of the season he should establish a winter base in Sining (modern Xining) so as to be able to explore northern Kansu the following spring.

This letter crossed with one written by Sargent on 30 August, addressed to Will in Peking and telling him that neither he nor Veitch wanted to extend his contract and that he should return to England after he had packed and despatched his collections from Peking (mis-typed by Sargent's secretary as St Petersburg, which can't have helped Will's morale when he received the letter).

Will, who of course knew nothing about this, set off northwards, to the giant plateau at the eastern end of the Qinling range, known as Mount Hua, an inaccessible series of peaks that were a centre for religious pilgrimage. The plateau was famous for the rare medicinal herbs growing there and, more importantly for Will, was thickly wooded, with pines and junipers predominating. Reginald Farrer, writing five years later, claimed that on this trip Will fell 15 feet onto rocks when a metal ladder on the path up the mountain failed, and was left *"stunned and helpless"*.[80] Fortunately, according to Farrer, Will was found and nursed back to health by a Daoist nun, one of the hermits who lived in the foothills of Lianhua Feng (Lotus Flower Peak).

Reformed robber chief Acho-che-ro who befriended Will and made him gift of a long sword with which to defend himself (it may also have been a local status symbol).

Arnold Arboretum Archives

---

* Will, Sargent and Harry Veitch all knew that it would have been tantamount to suicide to travel to Labrang without the consent of the lamas who, despite Buddhist strictures to the contrary, did not hesitate to use violence, up to and including lethal force, to maintain their authority.

CHAPTER 10

Sargent received Will's letter in early September and forwarded a copy to Harry Veitch with a note telling him that he had written to Will saying that his contract would not be extended. Veitch replied that he quite agreed: his nephew's health had taken a turn for the worse and he had been ordered to cease work for three to six months. Even if John Gould Veitch made a partial recovery,* it was clear that he would never be able to run the family business. The lease on Coombe Wood expired in four and a half years; Harry Veitch, being then 71 years old, was not inclined to set up a new nursery and was trying to reduce expenses as much as possible. Finally, *"if we continued Purdom for another year we should not have time to see some of his plants show their true character before the Coombe Wood Nursery ceases to exist"*. In short, he agreed that Will should return at the end of the present collecting season. Veitch concluded by saying that although little that Will had sent so far seemed very promising, he hoped there might be *"something good"* in the current season.

Very shortly after this exchange, the 1911 Chinese Republican Revolution, also known as the Xinhai Revolution, broke out. *"A ragged affair that caught fire province by province"*,[81] which involved disparate groups of revolutionaries with different motives and aspirations, it nonetheless brought about the fall of the Ch'ing dynasty and the end of Manchu rule and destroyed the 2,500-year-old system of government of the Celestial Kingdom.

The Revolution is usually said to have started on 9 October when a bomb being put together by a revolutionary group in Hanchow (now Hankou) detonated prematurely. The revolutionaries, who included many ethnic Han soldiers, were forced to choose on the instant between fight and flight. They fought, and after a pitched battle in which several hundred Imperial loyalists and revolutionaries were killed, they captured the local government headquarters and the armoury. Only then did they appoint a leader, their brigade commander, Li Yuanhong.

Ten days later, Han troops, in alliance with the local Muslim gentry, massacred at least 10,000 ethnic Manchus in Xi'an† and took control of Shensi province. The Provincial armouries were broken open and the contents distributed on a 'first come, first served' basis, leading to a proliferation of armed gangs determined to derive personal advantage from the anarchy in the countryside, and making travel in rural areas very hazardous indeed. Two days after that, rebel soldiers overran Hunan province. By the end of October, seven provinces south of the Yangtse river, the most prosperous and populous part of China, were controlled by local alliances of soldiers, secret societies, local merchants and political movements. By the end of November, 11 provinces were in a state of mutiny and in early December government troops

---

* As his doctors probably anticipated, he did not recover but instead continued to deteriorate, to the point that when, in 1914, Harry Veitch wrote to Sargent with news of John Gould's death, he told Sargent that this was the very best thing that anyone who cared for John Gould could have hoped for.
† The *Times* correspondent in Peking, Ernest Morrison, put the number of Manchu dead in X'ian at 20,000.

evacuated Shanghai and Nanking without a fight. On 1 January 1912, a republic of 17 (out of 22) provinces was proclaimed, with Nanking as capital and the leader of the largest Chinese political movement, Sun Yat Sen, as President.

All over China, the authority of the Imperial government collapsed as regional officials melted away, joined the rebels, or were killed. The *ad hoc* local administrations which replaced the Imperial authorities were for the most part based on alliances between local gentry and the local military. They exercised tenuous control in the principal towns but not in the countryside.

In all this, the only good news for foreigners in China was that both the revolutionaries and the Imperial government used their best endeavours to protect them, partly for fear of reprisals from foreign powers and partly because both sides wanted Western governments to allow them to procure weapons from abroad and secure loans from Western banks to pay for them. But in the confusion and rapine in Xi'an, eight foreigners, six of them children, were killed. The revolutionaries in charge at Xi'an immediately issued a proclamation that foreigners should not be harmed and reinforced the message by publicly beheading the gang responsible for the murders,[82] but a party of foreigners which subsequently attempted to travel from Xi'an to Shanghai was ambushed after less than 20 miles, badly beaten, and driven back into the relative safety of the city.

The ability of the Ch'ing regime to fight the Revolution was severely compromised by the Regent's 1909 decision to inform Yuan Shih K'ai, the most effective general in the Imperial armed forces, that due to his poor state of health (of which Yuan had previously been quite unaware) he was relieved of his command and permitted to retire to his home province of Henan. Yuan was a leader of men who was strongly committed to modernising the Chinese army, particularly the 65,000-strong Beiyang Army, which he commanded and whose officers' first loyalty was to Yuan personally. He was also assiduous in cultivating good relations with foreign diplomats in Peking, who reported to their capitals that Yuan was the strong leader they considered essential to restoring order to China.

When news of events in Xi'an reached Peking, Sir Alexander Hosie and Sir John Jordan agreed with John Charles Keyte, an English Baptist missionary and Arthur de Carle Sowerby, a naturalist, explorer and writer, that Sowerby should lead a small armed group to Xi'an where the remaining missionaries in Shensi province were trapped, and would then escort them back to Peking. Hosie and Jordan were unenthusiastic about this initiative because they feared the group would get into a firefight with soldiers from one side or the other, thus compromising Britain's policy of strict neutrality. But as Keyte, whom they considered to be the loosest of loose cannons, was determined to go, they put Sowerby – 'a safe pair of hands' – in charge.

Sowerby was thoroughly familiar with western China and at the time the Revolution broke out he had been on the verge of setting out on a hunting expedition to Mongolia. He put the supplies, arms, and ammunition he had assembled at the

CHAPTER 10

disposal of what he vaingloriously called 'the Shensi Relief Column', a group of seven mounted volunteers of which he was the leader. The group left Taiyuan on 4 December 1911 and after several white-knuckle encounters with heavily armed groups of bandits or revolutionaries, all of which fortunately passed off without violence, they reached Xi'an on 27 December. They gathered together the 31 foreigners in the city and a large number of Chinese Christians and others who wanted to reach Peking, and left the city on 4 January, heading due east to the railhead at Honan, picking up more foreigners and Chinese followers on the way. After further hair-raising encounters with the Shensi Revolutionary Army and the Imperial army, which was preparing to attack the revolutionaries, Sowerby was able to persuade both sides to agree a ceasefire to allow the Westerners and their followers to pass through their lines. In mid-January they reached Honan, where they boarded a special train chartered for them by Yuan Shih K'ai and reached Peking on 17 January.

Will, meanwhile, was in Minchow, 300 miles west of Xi'an, where he was cut off from the outside world, hoping that before too long it would be safe for him to return to Peking. With characteristic sang-froid, in late October he travelled to the Peling mountain range west of Minchow, just over the Tibetan border, where he gathered a large quantity of seeds from several dozen tree and plant species.

This trip to the Peling range was not quite as hazardous as it might seem, since Will was in the company of two English big-game hunters whom he had befriended in Choni. George Fenwick-Owen, who had organised and funded the hunting safari, was collecting animal specimens for the British Museum of Natural History. He was also interested in plants and had collected a fine peony from the monastery gardens at Choni and a handsome clematis. He and his friend Harold Wallace had travelled from Britain to hunt in the mountains of northern Kansu, where the wildlife included wolf, leopard, bear, big-horned sheep, several different species of deer, and the takin, *Budorcas taxicolor*, a strange animal resembling nothing so much as a wildebeeste covered with wool. They had engaged as a guide Dr. Jack Smith, a former missionary who was thoroughly familiar with the region and who had in 1909 and 1910 been a member of the Bedford expedition.

Jack Smith and the Chinese staff of the party had made quite sure the locals knew that the hunters were first-class shots and possessed several rifles in different calibres and a shotgun each. They were not disturbed during the three weeks they spent camping in the Peling range. While Will collected seeds,[*] Fenwick-Owen and Wallace bagged a bear, two deer, and a big-horned sheep. They also found an animal new to Western science, the Kansu mole.[†] Then, on 11 November, a messenger from William Christie reached them summoning them urgently to return to Minchow.

---

[*] He may also have done some shooting: in later years he enjoyed 'walking up' birds with a shotgun and he also tried his hand, unsuccessfully, at stalking big-horned sheep.

[†] In due course it was given the scientific name *Scapanulus oweni*.

George Fenwick-Owen and Harold Wallace departing Lanchow for England via Moscow, mid-November 1911. Royal Geographic Society

When they learned about the killings in Xi'an, Fenwick-Owen and Wallace decided that it was not possible for them to continue their Chinese safari, and that they should return home via the only route then open, due north through Lanchow (modern Lanzou) to Omsk, thence to Moscow. Will took a photograph of their dawn departure from Minchow, and after a gruelling crossing of the Gobi desert they reached Moscow safely.[83]

It is apparent from a letter Will wrote to Morrison on 1 December that he had received Sargent's letter telling him that he and Veitch did not want to extend his contract, but it was impossible to hire mules and no one was travelling because of bands of robbers on the roads. The official in charge at Minchow was doing his best to stop the growing unrest in the town but it was an uphill task and *"a rising is daily expected"*. As soon as he could obtain transport he would 'come in to Peking' before returning home. Meanwhile, opium was once again being cultivated and consumed in the region. Typically, Will expressed concern for the effect this would have on the people of Kansu: *"it will take years to get over* [this] *move."*

Yuan Shih K'ai now found himself being implored by the government to quit his residence in Henan which, in a rare flash of humour, he had named '*The Garden of Increased Longevity*'* and to lead the fight against the revolutionaries. He drove

---

* The joke being that the name was simultaneously a trope of the Confucian ideal of the gentleman scholar living in serene harmony with the natural order of the Universe and a sardonic message to the Regent that (in 1909) Yuan understood that, if he wanted to keep his head on his shoulders, he should stay in his garden and keep out of politics.

CHAPTER 10

Lanchow, mid-December 1911: the escort of Muslim soldiers provided by the authorities to enable Will to return to Peking. Lakeland Horticultural Society

a hard bargain, demanding command of all Imperial forces and a commitment to political reform. The government eventually agreed, and Yuan led the Beiyang Army down the railway from Tientsin to Wuhan, where he took and burned the city of Hanchow. Yuan was appointed Prime Minister on 7 November 1911, returned to Peking, and formed a government of ten Han Ministers and one Manchu. Yuan's former nemesis, the Regent, promptly resigned. Yuan opened a secret dialogue with Sun Yat Sen and together they drafted an act of abdication, which the Emperor signed on 12 February 1912. Sun Yat Sen then announced his intention to resign the Presidency of the new Republic of China in favour of Yuan, and did so on 10 March.

In London, Harry Veitch was increasingly concerned about Will's welfare, but on 31 January, four months after he had last heard from Will, he received from Will's sister Margaret a copy of a letter Will had sent her, evidently telling her that he hoped shortly to leave Minchow for Peking.

A fortnight later, Margaret called on Veitch at the Horticultural Club on Victoria Street, very near her London lodgings on Buckingham Palace Road, and showed him a letter from Will dated 3 December telling her that

"*at present I am in a bit of a fix, cannot get mules to carry goods, the roads are full of robbers and no one will travel […] In Minchow there are quite a lot of hooligans and we have been expecting them to rush the* [mission] *station but so far it is only talk.*"

1911: THINGS FALL APART

He feared he might be unable to get away for a while, which would be *"a great bother"* because he had a large quantity of seeds he had not been able to send off yet, but he was glad to say that he had finished collecting in the area.

Later in February, Harry Veitch received 23 packets of seeds from Will, which he had managed to send via Turkmenistan and Moscow. There was no letter accompanying the seeds, but Veitch wrote to Sargent to say that it appeared from the dates pencilled on the seed-packets that in November 1911 Will had been in good health and at work.

In the same month, Veitch received a letter dated 17 December from Will in which he reported that

"the situation here [Minchow] *is critical, people ready to rise at a moment's notice. The roads are blocked by mobs and robbers who murder and rob without the slightest heed. The local roughs threaten to kill me and my party but so far we have put up a bold front and have fair shelter, also some rifles."*

Will was, for once, giving his sponsors an unvarnished account of the dangers he was facing. The authorities were only just in control of the town itself and the adjacent roads and countryside were in a state of complete anarchy. To return to Peking Will would first have to head south for 100 miles, following the western edge of the great natural barrier of the Peling range, before turning east for over 200 miles through the mountains towards Xi'an, then a further 250 miles east, passing through an ill-defined demarcation zone between rebel- and government-controlled territory to the railhead at Honan, from where he could make his way north to Peking.

In normal times, the journey from Minchow to Honan would have taken a month or so, riding 20 miles a day and sleeping in rural inns and small villages and towns along the road. In late 1911, most of the inns and villages had been looted and burned and the roads were alive with armed bandits with no allegiance beyond a keen desire for self-enrichment. Will would have been assumed (rightly) to be carrying silver specie and other valuables, and would have been a prime target for robbers. His only sensible course of action was to stay where he was, but he told Harry Veitch that *"In due course I hope* [Minchow] *will be relieved by the reform party's soldiers then* [I can] *get an escort out of this part."*

This hope appears to have been realised very shortly after Will wrote to Veitch. Will's personal photo album from this period contains a photograph of four armed horsemen holding a fifth horse, presumably Will's, and labelled *"Moslem escort from Kansu, 1911"*. The accounts Will presented at the end of the expedition show that he spent $380 on *"escorts down country"* and $165 on *"feeding escorts' horses"*. His sister, Nell, in a letter written 65 years later,[84] said that her brother had employed *"a Mohammedan guard – a few armed men"* to escort him all or most of the way back to Peking.

As news of the Imperial abdication and the declaration of the Republic spread across China, the risk of being caught up in clashes between Imperial and

CHAPTER 10

Revolutionary forces faded away, but the armed escort still proved its worth. When he eventually reached the safety of Peking, Will wrote to Harry Veitch describing "*a very narrow escape*" when he was ambushed, probably in early March 1912, by brigands near Shenchow (now Zhengzou) in Honan province:

"*they attacked me with three Chinese escort, shooting down two of our horses** [but] *thanks to a rifle I had, a stand was made which turned the tables, and although I'm sorry to have had to shoot three Chinese with several horses to get clear, I feel pleased that this was the saving of my life.*"[85]

Will's habitual taciturnity irritated Veitch and Sargent, who would have liked to know more about this incident.† It appears that, after his return to Britain, Will gave Harry Veitch a more detailed account because two years later Veitch gave an interview to the *Daily Mail* in which he boasted of the "*free and necessary*" use of firearms by a Veitch collector "*escaping from China*". This can only have been Will. Unfortunately, the reporter did not see fit to record any further details.[86]

Aesculus chinensis in the Tsin Shih Tse temple in the Western Hills, Peking, photographed by Will's friend Joseph Hers in 1926. This was almost certainly where Will procured seedlings for Charles Sprague Sargent in 1914.   Joseph Hers/Arnold Arboretum Archives

---

* At least one of these appears to have been a packhorse, because Will recorded having lost some of his baggage and papers in the ambush, including the fair copy of the expedition accounts.

† Sir John Jordan may have encouraged Will to say as little as possible about it: the brigands could well have been off-duty government soldiers, making their deaths at the hands of a foreigner liable to be portrayed as an act of aggression against the Republic. On 10 March 1912, Ernest Morrison recorded in his diary that Sir John had shown him a telegram to London about the incident, but it has been 'weeded' from the Foreign Office file in the Public Record Office.

1911: THINGS FALL APART

When Will and his escort passed through Taiyuan, north of Honan, he found Mrs Soothill, the wife of Professor William Soothill, a distinguished scholar and Chinese linguist, and her daughter, Dorothea, staying at the local mission station. They were nervously contemplating the return journey to the capital, and were only too pleased to accept Will's invitation to join his little party.*

On 20 March Will arrived safely in Peking, and delivered Mrs Soothill and Dorothea to Sir Alexander Hosie, with whom they were staying. The 1,000-mile journey from Minchow had taken him three months, and he promptly sent Harry Veitch a telegram announcing his safe arrival. Will followed this up with a letter[87] in which he told him that "*my health is a bit run down through the struggle of getting out of Kansu, no food and shelter being at hand save scrap* [sic] *Chinese food*" but he hoped to "*pick up*" in a few days. He had lost "*some things*" in the ambush but had reached Peking with "*a fair amount of material*" and would be sending some more seeds very shortly. He noted that

Sir Alexander Hosie, Will, Mrs Soothill and her daughter Dorothea, in the grounds of the Temple of Heaven, Peking, April 1912. Dorothea had very recently agreed to marry Sir Alexander, which may account for the group's festive air, or were they celebrating Will's birthday?
Lakeland Horticultural Society

his contract had expired in February and promised to return home as rapidly as possible, but before leaving Peking he would secure the specimens of *Aesculus chinensis* that Sargent had particularly requested.

Will went to a good deal of trouble to find specimens of the *Aesculus,* which he knew Sargent was very anxious to obtain for the Arnold Arboretum. He visited and searched the grounds of some 40 temples before, on 30 April,[88] he found a

---

* Nell Purdom thought this encounter took place in X'ian, but this cannot be right. In her memoir, *Two Gentlemen of China*, Dorothea records that she and her mother were in Peking between January 1912 and at least early March. Furthermore, they had no reason to visit X'ian, and it would have been sheer folly in early 1912 to attempt to travel there from Peking. But the two women, who were about to leave China, did want to say goodbye to old friends in Taiyuan, where Professor Soothill had been President of the Chinese University and where Dorothea had set up a school for the daughters of the local gentry.

CHAPTER 10

grove of six fine trees in the Western Hills, where he paid the priests $25 to be allowed to dig up six seedlings small enough to be carried back to London in his luggage. Will was accompanied by one of his Chinese assistants whom he paid $30 to return in the autumn and collect seeds, which he delivered to the British Legation to be forwarded to Veitch & Co.

One fine day in April Sir Alexander Hosie, Mrs Soothill, Dorothea and Will together visited the grounds of the Temple of Heaven, where Will used the self-timer on his camera to photograph all four of them. He also took portraits of himself and Sir Alexander. There is a festive air about the photographs, which may reflect the fact that around this time Dorothea Soothill and Sir Alexander became engaged, or they may all have been celebrating Will's 32nd birthday, on 10 April 1912.

Will did not leave Peking for London until the first week of May 1912. This was not only because he was tidying up the loose ends of his three-year expedition. He was also seeking out interested Chinese officials and opinion-formers to make the case for effective management of Chinese forests and for reforestation, endeavours in which he was keen to be involved.[89]

During his travels in China Will had repeatedly come across the remnants of forests devastated by uncontrolled 'clear felling'. He was convinced that this destruction was directly linked to the recent catastrophic floods in China, because the lost trees had stabilised river banks and reduced soil run-off from the higher ground and because, in common with other scientists interested in forestry and the environment, he believed that large expanses of forest had a beneficial effect on local climate by reducing extremes of temperature and precipitation.

It appears that Will approached Ernest Morrison and lobbied him on these lines. There is amongst Morrison's papers an undated memorandum signed by Will entitled *'The afforestation question in China'*. The memo refers to Will's arboricultural research in the provinces of Chihli, Shansi, Shensi, and Kansu, so that it cannot have been written before his return to Peking in early 1912.* Morrison was meticulous about retaining and filing both inward letters and carbons of his replies to them, and the absence of a covering letter from Will or a letter of acknowledgement from Morrison strongly suggests that the memo was handed over during a face-to-face meeting in Peking between late March and early May 1912.

Will began by suggesting that a scheme of afforestation in China was urgently needed and "the time [is] *ripe for pointing out a few of the drawbacks attending the wanton destruction of what must have been fine old forests*". First amongst these was the serious impact of the loss of forests on the climate, erosion, and the silting up of rivers.

---

* The full text is at Appendix B.

> The memorandum recognised that fuel was hard to come by in rural China, but *"this does not excuse the wholesale scraping from mountain sides of every particle of vegetation [...] which results in the extinction of all soil-binding plants [...] This havoc is particularly galling when one thinks of the effect the development of the country's immense coal fields would have."*

Judging from the remnants of forest Will had seen in northern China *"excellent timber for almost any purpose must once have been available"*. He described in some detail the waste of good timber in the Weichang district of Chihli province and the chronic soil erosion around Jehol.

Will also set out the effect the lack of timber must have on the development of infrastructure including railways, mines and bridges, and on small-scale manufacturing. Finally, he recalled the encroachment of the Ordos desert south of the Great Wall in northern Shensi province, where planting of soil-binding plants would be of great benefit. He concluded by hoping that

> *"those who direct the agricultural policy of China will seriously consider the advisability of tackling the question* [of afforestation], *particularly as China contains within itself adequate and excellent stock* [of tree species] *from which to make a beginning."*

Morrison does not appear to have taken any action on the memorandum beyond filing it away for possible future use. He may have believed that against the background of the chaotic and deeply divided state of the country there was no prospect of getting the Chinese government to focus on these issues. If so, he was mistaken: one young official in the Ministry of Agriculture was doing just that.

Aged 29, Han An (in pinyin, Han Ngen), the Secretary of the Minister of Agriculture, was typical of the young men and women whom the Chinese government had for some years been sending abroad to study technical subjects so that on their return to China they could help drive forward the modernisation of the country. Han's father was a labourer; his mother worked in a mission station in Anhui Province, and a perk of the job was that Han could attend the mission school without charge. From there he went to Nanking University and in 1907 he won a Chinese government scholarship to Cornell, and then to study forestry at the Universities of Michigan and Wisconsin. He returned to China in January 1912 and was appointed to the Ministry of Agriculture, where he started to put together a proposal for a Chinese Forestry Department

There is no record of how and when Will and Han An first met. It is possible that the British Legation was involved. Will's friend, Sir Alexander Hosie, was responsible for promoting British commercial interests in China. His engagement with Chinese botany and botanists meant that he would have known about discussions going on within the Ministry of Agriculture in early 1912. Will possessed precisely the skill-set the proposed Forestry Department would need, and it would have been natural for Hosie to introduce him to Han.

CHAPTER 10

What is clear is that the two men met in Peking in the spring of 1912, after which Will ceased to advance the general argument set out in the memorandum he had given Morrison that 'something must be done' about Chinese afforestation, and instead pushed for the creation of a Chinese Forestry Department, as proposed by Han. I believe that by the time he left Peking in May 1912 Will had begun a dialogue with Han which continued after his return to Britain and led him to hope that in the not-too-distant future he might return to China and work to protect the nation's existing forests and to restore those devastated by uncontrolled logging.

Chapter 11
The first expedition: a reckoning

Will arrived in London on Sunday, 19 May 1912. The following day, he called on Harry Veitch in Chelsea and gave him ten primulas, three hypericums, two aconites, an abelia, a polygonum, a plant tentatively identified by Veitch as *Zizania latifolia*, Manchurian water rice, and six *Aesculus* seedlings, all of which he had brought back in his luggage on the trans-Siberian express (he had experienced some difficulty keeping the plants alive in the overheated carriage).

Will also handed over the balance of the petty cash from the expedition funds and his accounts for 1911/12, which he had reconstituted from his pocket-book. The total cost of the expedition was a little more than $5,500 (£1,200) and in the course of it Will had travelled over 8,000 miles around China, mostly on foot or on horseback.

Will gave Harry Veitch a detailed account of his adventures on the road between Minchow and Peking, and they discussed the results of the expedition, which the sponsors considered to be disappointing. Although Veitch's overall assessment was that "*if the trees and plants weren't there, then he* [Will] *couldn't send them*",[90] both Veitch and Sargent felt that Will had shown himself incapable of working to the same standard as that achieved by Wilson, and Will could not deny that he had failed to find more than a small number of new plants.

Later the same day, Will wrote to Sargent reporting his meeting with Veitch, and that he had arranged for one of his former employees in Peking to send him *Aesculus* seeds in the autumn.

Will would come to believe that he had not been fairly treated by either Veitch or Sargent and that before leaving for China he should have been given more extensive training, especially in the preparation of herbarium specimens. But at this time he was apologetic about the paucity of new species amongst the plants and seeds he had collected, which he attributed to the plants in his collecting-area being substantially identical to those in the area in which Wilson had previously collected.*

Will told Veitch that "*in view of the unfortunate situation the expedition has been placed in and the absence of novelties*" he did not expect to be paid for the period after the term of his contract, the end of February 1912. Harry Veitch, a fair-minded man, wrote to Sargent that although Will was being

---
* We now know that Will was right to believe that there were almost no plants in his collecting-area which were not also present in Wilson's, but quite wrong to believe that the converse was also true.

CHAPTER 11

"straightforward" about his salary, since he had been detained in China past the end of February he should be paid until the date of his return to London, and Sargent concurred.

Harry Veitch also wrote to Sargent in June and again in September 1912, promising him two *Aesculus* seedlings. Although there is no record of their despatch or arrival in Boston, *Plantae Wilsonianae,** the comprehensive catalogue of plants collected from China for the Arnold Arboretum published in May 1913, records that *"Aesculus chinensis has recently been reintroduced* [to the West] *by the Arnold Arboretum through its collector W. Purdom."*[91] Veitch may have sent seedlings to Boston in late autumn, when they would have been dormant and thus better able to survive the journey: he certainly presented one to Kew in December 1912. The Kew seedling thrived but the Boston ones appear to have died, as did two further seedlings procured by the Arboretum in December 1913 and December 1916.† As late as 1933, the Arboretum lacked a specimen of *A. chinensis*.

All plant material sent by Will from China during the course of the 1909–12 expedition was addressed to the principal offices of James Veitch & Sons in Chelsea, where it was logged on arrival in a register labelled 'Seeds collected by W Purdom' (in fact, all plant material, not just seeds, was entered in this book). The bulk of the material was promptly put on the daily delivery run between Chelsea and the Veitch nursery at Coombe Wood. Packets of seeds were sometimes divided first so that half could be sent to Boston and occasionally individual plants or descriptions of plants were sent to Kew for identification.‡

The plants sent to Coombe Wood were entered into the substantial ledger of 'Purdom Collections' which the nursery had maintained since 17 May 1909, the day the first plants had arrived from Will in China. Harry Veitch sent to Coombe Wood the plants Will handed over at Chelsea on 20 May 1912, where they arrived and were entered in the ledger the same day. This formally concluded the 1909–1912 Purdom expedition.§

Tragically, the Veitch archives, which spanned over a century during which the firm was at the summit of the British horticultural industry, were put on a bonfire in the 1950s. By a remarkable coincidence, however, both the Chelsea 'Record of Seeds' sent by Will from China and the Coombe Wood ledger of 'Purdom Collections' escaped the flames, the former having been given in July 1914 by

---

* Notwithstanding the title, not all the plants listed in the compilation were collected by Wilson.

† One seedling (AA plant ID 10662*A) was purchased in December 1913 from the German nursery Messrs Hesse of Ems and the second one (AA plant ID 12668*A) was sent to the Arboretum by Kew.

‡ For many years, anyone could send a specimen of a plant to Kew for identification, without charge. The only *quid pro quo* was that polite requests from Kew for seeds, cuttings or herbarium specimens of those plants sent in which were of interest to the Royal Botanical Gardens were hardly ever refused.

§ Any reader who takes the trouble to look at the ledger in the Lindley Library archive room will immediately spot that the very last entry in fact relates to a specimen of *Ulmus pumila 'pendula'*, Peking elm, which is recorded as arriving in Coombe Wood on 11 June 1912. This is a clerical error: the tree is logged as having reached Chelsea on 1 May and was despatched the same day to Coombe Wood, where it appears no-one thought to enter it in the register until six weeks later.

Harry Veitch to Kew and the latter[92] gifted in 1920 to Fred Crittenden, the first Director of the Royal Horticultural Society garden at Wisley, Surrey.

The Chelsea record book lists every box, parcel or packet received from Will at the head office of Veitch & Co, and the date on which it arrived, including parcels of seeds and germplasm sent by the Arnold Arboretum as part of the agreement to share Will's collections.* It also comprises a list of Will's collection numbers received by Veitch & Co and identifies the plant concerned, what the material consisted of, and where it was sent from Chelsea (usually to Coombe Wood, but some material went to the Veitch nursery specialising in alpines and fruit trees at Langley, and some dried specimens were retained at Chelsea). There are some blanks where it appears material was lost in transit between China and Britain.

The Coombe Wood ledger lists every item of plant material received there. It records the specimen's date of arrival at Coombe Wood; where appropriate, the date sown; a description or identification of each item and, usually, details of where and when the plant or seed was collected.

In short, the two Veitch volumes between them give a comprehensive overview of the plants collected by the 1909–12 expedition. The equivalent records of the Arnold Arboretum listing the specimens received from Wilson in 1910/11 also survive, making it possible to reach an informed judgement about Sargent's criticism of Will by reference to Wilson's alleged greatly superior skills as a collector.

The basic figures are easily set out. Following standard practice, Will numbered sequentially every specimen he collected. No. 1 was an anemone sent from Peking in April 1909, and the series ended with no. 871, a *Pyrus* (pear) from south-west of the Tao river that arrived at Coombe Wood in April 1912. Neither the Chelsea list nor the Coombe Wood ledger records the collection numbers of the plants Will delivered personally to Harry Veitch in May 1912, but they bring the total sequence to 891.

During the 1910/11 collecting season (defined as from 1 April 1910 to 31 March 1911), the first specimens to arrive in Chelsea from Will were numbers 305 and 306. The last of the season, no. 679, a *Prunus*, arrived on 22 February 1911. So, during the 1910/11 season Will sent specimens or elements of 374 different plants to Chelsea,† including several instances where a single collection number covered up to four separate packets of seeds or several different elements from

---

* A good deal more material travelled east–west across the Atlantic than vice versa. Once Will realised how much longer plant material took to reach Sargent, he favoured Veitch, especially for seeds, which had a better chance of reaching Chelsea in a viable state. They could then be germinated, and plants or the next season's seeds forwarded to the Arnold Arboretum. The flow of material from London to Boston is recorded in letters between Harry Veitch and Charles Sprague Sargent and in the Chelsea register but does not figure in the Coombe Wood ledger, whereas material sent by Sargent to Veitch is recorded in both the Chelsea register and the Coombe Wood ledger.

† In theory, this total might include a number of duplicates, the same plants inadvertently registered under more than one collection number. Exactly the same is true of Wilson and his collections. But both men were highly competent at identifying plants and there are likely to have been few such instances, if any.

CHAPTER 11

the same plant, each in its own wrapper or packet, for example foliage, seeds, and flowers.

Over the same season, Wilson sent back to Boston specimens identified by 744 collection numbers, 281 of them attributed to his Chinese assistants, working to Wilson's instructions after he had broken his leg.

Accordingly, over the 1910/11 season Wilson and his team collected specimens from almost exactly double the number of plants which Will and his (much smaller) team sent during the same period.

Sargent's observations concerning the work of Will Purdom and Ernest Wilson in the Arnold Arboretum Director's *Annual Report to the President of Harvard University, 1910–11*, are at first sight difficult to reconcile with the above. Sargent said a good deal more about Wilson (*"the results of his second journey are the seeds of 462 species of seeds and shrubs [...] 2,500 sheets of herbarium specimens and 374 photographs"*) than about Will (*"Good results have been obtained from Mr Purdom's second season in China"*) and concluded that *"On his [Wilson's] last expedition he sent back 1,285 packets of seeds [...] In an equal period Purdom sent only 304"*.

On the face of it, Sargent was saying that Wilson found and sent to Boston rather more than four times as many plants as did Will over the same period. Clearly, this is quite wrong, but how and why did the mistake arise?

Part of the answer to the first question lies in the precise language used by Sargent, specifically in his reference to 'packets' of seeds. Wilson was notorious within the community of those interested in Chinese plants for collecting and sending to his sponsors very large quantities of seed from each and every plant he came across, regardless of whether they were novelties or were already widely cultivated in the West.* When Wilson was working for Veitch and Co, Harry Veitch wrote to him several times about this. In February 1901, for example, Harry Veitch – who during the course of the previous year had received 1,000 packets of seeds, many of them duplicates or bundles of up to ten identical packets of seeds – told Wilson to send *"not too much seed of ordinary things; and even of the very best three or four packets is enough."*[93] Nonetheless, Wilson more than once sent such large quantities of seed that the surplus had to be dumped on the rubbish-tip at Coombe Wood.[94]

It is, accordingly, entirely possible that in 1910 Wilson sent Sargent over a thousand packets of seeds. But the impression created by referencing packets rather than species is to overstate Wilson's 1910 harvest, especially in contrast to Will's. Sargent, a highly intelligent man thoroughly familiar with the practice and methodology of plant-hunting expeditions, must have known what he was doing,

---

* For example, when in 1917 Reginald Farrer was urging Augustus Bowles to sponsor a proposed post-war plant-hunting expedition he promised that the plants sponsors would receive from China and Tibet would be choice specimens, "<u>not</u> *Wilsonian weeds*".

which was deliberately causing Wilson to appear to be an almost preternaturally capable collector, in contrast to the apparently incompetent Purdom. Leaving aside for a moment the grave injustice to Will thus perpetrated, why did Sargent do it?

The answer may well lie in the constant struggle to raise funds in which Sargent was engaged during the first decades of the 20th century. The will of James Arnold (*d*.1868) provided an endowment which funded the Arboretum's core operations. But Sargent always needed more money to enable the Arboretum to grow its estate and expand its activities. He was a brilliantly successful fundraiser and the Arboretum today – an institution that stands amongst the very best in the world in the field of scientific botany whilst simultaneously managing a remarkably fine arboretum and botanical garden much loved by the people of Boston – is a monument to his ruthless determination (and, it must be said, to the generosity of the 'Boston Brahmins' whose chequebooks Sargent sometimes came close to treating as his personal property). All this was long before the days of such jargon as 'centre of excellence' and 'unique selling point', but Sargent, like any modern professional fundraiser, well understood that to maximise donations an appeal for funding must stand out from the rest.

Sargent repeatedly emphasised to the President of Harvard that the thousands of plants sent back from China by Wilson, many of them new to Western science, made the Arboretum the pre-eminent global holder of Chinese flora. He argued passionately that the University should provide at least some funding to enable the study, classification and exploitation of these holdings,* and he deployed exactly the same arguments when appealing to potential private donors, who often proved more persuadable than the Harvard Board of Overseers. In plain English, by painting Wilson as a superhero of plant-collecting Sargent was amassing credit, both literally and figuratively, for the Arboretum.

Sargent's praise for Wilson's collections in China should also be seen in the context of the hero's reception that greeted Wilson when he returned to Boston in March 1911. His account of how, after his leg had been smashed in an avalanche, the only way he could get past a mule-train going the other way along a narrow path with sheer rock on one side and a 500-foot drop on the other was to lie motionless across the track whilst the animals stepped over him, one by one, captured the imagination of the national media and was widely reported. He found a ready market for the articles and books which Sargent encouraged him to write about his expeditions, and in which Wilson stressed his links with the Arboretum and with Harvard. Unsurprisingly, these best-selling accounts of his adventures are very much written in the first person. Whilst there is no reason to believe that Wilson felt any hostility towards Will, it is clear that he did not wish to share

---

* *i.e.* funds for immediate use, what might be called current expenditure.

any of the glory being dispensed by the American and British press with Will or with anyone else. In short, Wilson's and Sargent's interests coincided, and jointly worked to the detriment of Will's reputation as a plant-hunter.

Nothing in the foregoing seeks to challenge the established consensus in the community of historians and students of plant-hunting in the late 19th and early 20th centuries that Wilson's expeditions to China were brilliantly successful and that, should it be considered to be useful to rank the collectors of this period in order of merit, Wilson is a strong candidate for first place. But what *is* suggested is that Wilson's marvellous harvest of plants from China was partly enabled by the wholly exceptional botanical wealth of the region in which he collected, and that his heroic status in the plant-hunter's pantheon derives in part from his being 'taken up' on his return from China by the US and international press. Furthermore, the stellar reputation achieved by Wilson unfairly eclipsed the achievements of several of his peers, including Will Purdom and Frank Meyer.

In the light of the above, what objective conclusions can be drawn concerning Will's first expedition to China?

First, he did well to come home at all. In late 1911, by sheer bad luck, Will was collecting in the region of China most violently disrupted by the Xinhai Revolution. He survived because he kept a cool head when faced with immediate threats to life and limb and also because – unlike other collectors, including Wilson – he had taken the trouble to learn to speak Chinese and instinctively and unhesitatingly rejected the dyed-in-the-wool racist attitudes endemic in the expat community in China. If Will had not consistently made friends in the communities local to where he was collecting, people who could, if they chose, offer advice and protection, it is doubtful whether he would have survived.

Second, Will consistently displayed exemplary commitment to the task in hand, often in the face of considerable hardship and difficulties not of his making. His collecting at all in the autumn of 1911, for example, rather than staying put in the relative safety of Minchow, reflected professional engagement of the very highest order.

Third, whilst Sargent's 1909 decision to send Will to collect in the botanical *terra incognita* of north-west China, the Chinese/Tibetan end of the Qinling mountain range, was an entirely reasonable throw of the dice, his subsequent refusal to recognise that the disappointing outcome of the gamble was because, to quote Harry Veitch, "*the plants were not there*" was, at best, wrong-headed. Insofar as it amounted to a refusal to accept responsibility for what was unequivocally his (Sargent's) personal decision, it was blatantly unfair.

Fourth, if one looks at the quantity and quality of the material Will sent back from China with the inestimable advantage of detailed modern knowledge about the area in which he was collecting, it appears that he did rather better than his sponsors could in reality have expected. The bottom line is that collecting in an area in which there grow between one-third and one-eighth of the number of different plants found in the region in which Wilson was working, Will sent back in the course of one collecting season specimens of half as many plants as did Wilson during the identical season. This crude measurement of the relative efficiency of both collectors should definitely not be taken too far, but it does put their work into perspective. There is no definitive list of the number of plant species sent by Will to Boston between 1909 and 1912 but his final tally of seeds was 550 packets, plus other plant material from which in his lifetime specimens of 143 different species were transferred into the Arboretum. He also sent to the Arboretum over 1,100 herbarium specimens – objectively, a good haul.

Fifth, it must be recognised that Will's taciturnity did him no favours with his sponsors. He was, as Harry Veitch observed, a poor correspondent.[95] His sponsors' failure, for most of the duration of the expedition, to grasp the difficulties let alone the dangers encountered by Will in China was largely due to this reticence.

Finally, what can we conclude about his work for the British secret services? The answer is very little: it was, after all, secret. What we do know is that Will engaged in and supported intelligence-gathering at the request of interested parties in the British Legation, probably involving map-making, which was assessed by those concerned as valuable and helpful to the interests of His Majesty's Government. It's unlikely that Will sought or received anything beyond private thanks for this service, but it is certain that on his departure from China a discreet note was made in the Legation files, in case he should one day return.

Chapter 12
Loose ends in London and return to Westmorland

Will had returned to London at the beginning of what was a very busy time for Harry Veitch. He had for many years been involved in the organisation of the Royal Horticultural Society (RHS) Great Spring Show, which was held in May at Temple Gardens, between Fleet Street and the Embankment in London. In 1910 it became apparent that the show had outgrown the Temple site, and a committee, of which Harry Veitch was a leading member, was set up to organise the 1912 'Royal International Horticultural Exhibition' in the grounds of the Royal Chelsea Hospital.*

Will is said to have been asked by Harry Veitch to support the James Veitch & Sons stand at the week-long Exhibition, which was opened by King George V on Wednesday 22 May. The Exhibition was primarily a showcase for British horticulture and most of the stands were displays by British nurserymen or leading British amateur gardeners, but 20 other countries also had pavilions showing their native flora. Many plants introduced from China by Ernest Wilson were in evidence, including a large display of shrubs and trees grown from seed sent by Charles Sprague Sargent to the prominent Hertfordshire gardener Vicary Gibbs, whose stand won a Gold medal.

The Veitch stand showed over 150 species of Chinese plants described as collected by Wilson for the firm, and won another Gold medal. Ellen Willmott of Warley Place, Essex, won a silver cup for a display of Chinese herbaceous plants raised from seed collected by Wilson. His Majesty was impressed, and announced in his speech at the opening ceremony that, in recognition of a lifetime of service to horticulture and horticultural charities, Harry Veitch was to be knighted.†

There are no references in the detailed reports of the Exhibition in the *Gardeners' Chronicle* to any of Will's collections, any more than there had been any mention of him in the 1911 and 1912 Veitch catalogues of *New Hardy Plants from Western China (introduced through Mr. Ernest Wilson)*.

---

* For non-British readers, the Chelsea Hospital was founded by Charles II in 1681, as a 'retreat' (i.e. a retirement home, the original meaning of the word 'hospital') for retired British soldiers in need. It occupies a 66-acre site between Sloane Square and the Thames. Paradoxically, it is now best-known as the venue for the annual Royal Horticultural Society Chelsea Flower Show.

† The King's enthusiasm for the Exhibition caused him to jump the gun: when the knighthood was gazetted (formally announced) on 28 May, it was necessary for the notice to stipulate that it was retrospective to the 22nd.

CHAPTER 12

This silence concerning Will was partly because although Harry Veitch had sold some of Will's seeds and plants to special customers, including the tree and peony enthusiast Robert Woodward, there had been insufficient time for the firm to 'grow on' Will's introductions in the quantities required to bring them to a wider market.* But Veitch's failure publicly to recognise Will's efforts may also have derived in part from his perception that although Will was a skilled worker, capable of getting the job done, credit for the expedition belonged to the 'gentlemen' who had been its sponsors.

Whatever motive Will attributed to Harry Veitch, he must have been disappointed that his efforts over three years were passed over, not least because good publicity could have been helpful to his search for new employment. Some indication of Will's state of mind at this time can be inferred from his failure to attend the 1912 annual dinner of the Kew Guild, held on 28 May, when over 100 Guild members met at the Holborn Restaurant in London.

This decision probably reflected a disinclination on Will's part to rub shoulders, along with a number of old friends, with senior staff members of less happy memory, including Colonel Prain. It may also have been the case that on his return home the considerable stress Will had been under for most of the previous year caught up with him and he experienced what would all too soon come to be known as 'shell shock' or, in today's parlance, post-traumatic stress disorder (PTSD). We have few details of the ambush in which, less than four months earlier, Will had been forced to shoot three Chinese brigands and several horses, but it is unlikely that this sensitive and introverted man simply shrugged off an experience which at the time presented a strong possibility he would be murdered and which must in hindsight have caused him to examine his own conscience about whether there had been any alternative to killing three men. In any event, although he called at Kew,[96] Will did not attend the dinner.†

On 6 June, Will and Sir Harry went together to Coombe Wood to see and to talk over his introductions, after which Will travelled to Ambleside to spend a fortnight with his family and friends there. He then returned to London and spent some time helping Sir Harry as he passed on to the Royal Horticultural Society the details of the organisation of the Exhibition in order to enable the Society in 1913 to arrange the first RHS Chelsea Flower Show. After this, Will's connection with the firm and with Sir Harry ended and by the autumn of 1912 he had moved from London to his parents' home in the Lake District.

Little is known about Will's life at this time. The planned marriage to Annie Groombridge, now 23 years old and earning her own living as an elementary school teacher, did not take place. There is no clue as to whether it was Annie or Will

---

* This process could take five years or more: for example, Veitch & Co used the Exhibition to launch two new roses, *R. Moyessi* and *R. Willmottiae*, which had been collected by Ernest Wilson in 1903 and 1904.

† Will didn't attend the 1913 dinner either.

who broke off the engagement. Will is known to have written to her and to have sent small gifts from China, but his peregrinations mean that his letters are likely to have been infrequent, and her opportunities to reply even more so. Perhaps they simply discovered that during their three-year separation they had grown apart. The fact that Will was out of work, with no immediate prospect of finding employment, cannot have helped.

From Ambleside, Will wrote a short article, illustrated with two photographs, for the *Gardeners' Chronicle* on the subject of *Aesculus Chinensis*, in which he described in detail the specimens he had encountered in Peking, quoted Sir Alexander Hosie's explanation of the etymology of the tree's Chinese name, and described the use of the nuts in traditional Chinese medicine to treat stomach ache.

Frank Meyer, the Dutch-American plant-hunter who came to Britain in November 1912 specifically to seek Will's advice on collecting in north-western China.
Frank Meyer/Arnold Arboretum Archives

In November, Will received a visit from the Dutch-American plant-hunter Frank Meyer, who had spent the period between late 1909 and October 1911 collecting seeds for the US Department of Agriculture in Russia, Georgia and the Caucasus, thence into Chinese Turkestan and Mongolia. Meyer had hoped to conclude his expedition by going to north-west China in search of Chinese persimmons, but reports of violent unrest in Shansi and Kansu provinces had caused him to cut his expedition short and to return, via Europe, to Washington.* After four months in America, including a fortnight at the Arnold Arboretum during which he had mended his fences with Wilson, Meyer was once again bound for Kansu and Shensi provinces.

Sargent pointed out to Meyer that Will was the only Western botanist to have collected in this area of China and, although unable to resist adding that Will had "*not done his level best to obtain certain things that come from there*", he suggested that Meyer should seek Will's advice about local conditions.[97] Meyer had accordingly routed himself via London and on his arrival there on 10 November he wrote to Sir Harry Veitch asking for Will's address. Sir Harry claimed not to

---

* Meyer, who in the course of his travels repeatedly encountered violence, disease, and brutal extremes of weather, was convinced that he benefitted from "*some sort of protective atmosphere*" surrounding his person. This belief was reinforced when, after having booked his passage to New York on the RMS *Titanic,* he fell ill and was forced to change his booking to the sailing, four days later, of the RMS *Mauretania*.

have it and advised Meyer to contact Will's sister Margaret, who lived in central London. She replied to Meyer's enquiry saying that her brother was "*out* [of London] *in the country*" but she had forwarded Meyer's letter.

Will, who was clearly not anxious to see Meyer, then wrote to him saying that he was in Westmorland and because of pressure of work did not have time for a meeting. Meyer was not impressed, especially since that week's edition of the *Gardeners' Chronicle* included Will's article on *Aesculus chinensis*.* He reasoned that if Will had time for such *"literary pursuits"* then he could make time for a meeting. Accordingly, Meyer replied to Will's letter by telegram: he was catching a train to Windermere, would make his way from there to Ambleside, and hoped Will would receive him.[98] Will could hardly refuse, and on 16 November he and Meyer spent several hours discussing plant-hunting in north-west China.

Immediately after returning to London, Meyer set off for Russia, and on 21 December he wrote from St Petersburg to his friend and employer, David Fairchild, the Director of the Office of Seed and Plant Introduction of the US Department of Agriculture.[99] Meyer's account of his meeting with Will deserves to be quoted in full:

"Re seeing Mr Purdom. Well, Mr Fairchild, I didn't tell you much about this, and why? for reason that there was not so much to tell about him. Imagine an officer who lost a battle and you have Purdom's case. – He went out for the Arnold Arboretum in cooperation with other parties. He was not given time to study up what had been done in North China. He had the difficult problem of making good in rather a poor region, while E.H. Wilson was pouring in hundreds of new plants from very rich regions, He was really not the sort of man who ought to have been sent out for such work because they didn't train him up for such work. Then he had lots of bad luck in connection with the Revolution in China, with baggage and collections going astray, with incompetent assistants in China and with non-interested propagators in England, and all this gives you Purdom's tale in brief.

I really pitied the poor fellow, just like I pitied myself often these first five years in China, where I had much the same battles to fight as he. Had he been somewhat more careful and attentive I think he would have done better; as it is now, Prof. Sargent didn't even invite him to come to Boston and give his explanation of it all. (This of course is really personal!)

He told me however what districts I ought to visit, and what time to go, etc., etc., but his misfortunes weighed so upon him that he rather did not wish to discuss everything in connection with his past long trip. Had I been able to take him out to "eine gemütliche Kneipe, vielleicht hätte er sein ganzes Herz ausgestürzt aber jetzt" Sie verstehen, nicht wahr, Herr Fairchild."

---

* The article concluded by mentioning that the photographs illustrating it *"are reproduced by permission of Sir Harry J. Veitch"*, from which Meyer may have concluded that Sir Harry's claim not to know Will's address was a deliberate untruth.

Meyer was born in the Netherlands and spoke fluent German. Before World War I, it was quite usual in the USA for members of the large community of Americans of German origin, and German speakers generally, to exchange private information in that language.

The phrase above translates as *"You'll understand, Mr Fairchild, that [had I been able to take him out to] a pleasant tavern, then perhaps he might have poured out his heart to me."* Clearly, the meaning Meyer wanted to convey was something he felt he needed to conceal beneath the cloak of a foreign language, but 100 years later it's difficult to be sure precisely what he *did* mean. The most plausible interpretation is that Meyer was emphasising the point he had already made to Fairchild (and would repeat in other letters)[100] that Will was so deeply depressed that he really didn't want to talk about his Chinese expedition and that he would benefit from getting out more. This in turn prompts the thought that the determined efforts by Sir Harry Veitch and Margaret Purdom to deter Meyer from meeting Will may have been prompted by concern that a meeting at this time might be bad for Will's emotional equilibrium.

What is clear is that by the autumn of 1912 Will had abandoned the semi-apologetic line about the results of his Chinese expedition which he had taken with Harry Veitch in May. It is also apparent that Meyer, an immensely experienced plant-collector who had just come from a fortnight in Boston largely spent in the company of Wilson and Sargent and who was thoroughly familiar with the background, agreed with Will that he had been inadequately prepared before the expedition and unfairly judged afterwards. There is a strong hint that the Veitch propagators at Coombe Wood may have unfairly prioritised Wilson's plant material, or perhaps they were simply swamped by the quantities of seeds and bulbs he sent them. Finally, Meyer's observation that Will might have fared better if he had been seen by Sargent to be more attentive to his concerns shows that he was not taking Will's word for what had occurred, but was exercising his own judgement and drawing on his own experience of dealing with Sargent.

All this, however, was in the past, and as Will drew down the savings he had built up whilst in China he must have been increasingly despondent about his future. His record as a trade union activist did not endear him to potential employers in the horticultural industry. It seems that Sir Harry Veitch, whose personal contacts in the British gardening community were second to none, was not inclined to help. There was a real prospect that Will might slide back down the greasy pole of social and economic advancement he had climbed with such effort over the previous decade, and end up working for a pittance as a gardener or market-gardener in some rural backwater. And, irrespective of the precise circumstances of the breaking off of his engagement to Annie Groombridge, it seems reasonable to suppose that Will may have been unhappy in his private as well as his professional life. Taking everything together, it is not surprising that he was depressed.

CHAPTER 12

White Craggs, Ambleside, the home of Will's friend Dr Charles Hough, whom Will helped to create an outstanding garden. Author's collection

Will filled in the time whilst casting about for permanent employment by working on the garden at White Craggs, a handsome Arts and Crafts house built in 1905 by Dr Charles Hough at Clappersgate, very near Ambleside. Mrs Hough, the daughter of Giles Redmayne, the owner of Brathay Hall, had died suddenly whilst Will was in China, and Dr Hough was determined to create a notable garden in her memory. He had known Will since he was a child and despite being 25 years his senior they became close friends.

White Craggs sits on a superb site, a steep hillside facing onto Lake Windermere incorporating a rocky crag which is a continuation of Loughrigg Fell. Mrs Hough had determined that instead of taking the (then) conventional route of either blasting the substantial outcrops of stone around the house flat or covering them with imported topsoil before creating a 'picturesque' landscape, she would expose the rocks and create a very large rock garden, using as many local plants as possible, so that the garden would blend into the magnificent adjoining scenery. In 1912, Dr Hough had barely started work, and Will entered with enthusiasm into the planning and design of what became an outstanding garden.* Will's involvement was so extensive that, writing in 1929, Charles Hough said that *"all that is of merit* [in the garden] *is due to his knowledge and foresight."* [101]

---

* The quality of the garden was such that in the 1960s the National Trust gave serious consideration to acquiring it, and only the absence of an endowment to fund maintenance deterred them from doing so. After a long period of neglect, it is currently (2020) being meticulously restored.

LOOSE ENDS IN LONDON AND RETURN TO WESTMORLAND

In the course of 1913, Will's morale improved, and he delivered a series of well-received lectures, illustrated with magic-lantern slides,[102] to various local groups about his travels in China and Tibet, and the manners and customs of the people he had met.* He also reminisced in private with old friends in Ambleside about his travels, when his affection for China and the Chinese people was very clear.

One hopes that Will went to London in May 1913 to attend the first Chelsea Flower Show, where Veitch & Co exhibited the small rose-violet *Aster flaccidus var. purdomii*, which received an Award of Merit. By the summer of 1913 he had certainly returned to London: he stayed with his sister Margaret, from whose address he wrote on 16 July to Ernest Morrison, who had the previous August resigned as the *Times* correspondent in Peking and been appointed a Special Adviser to the Chinese Presidency.[103] In 1912, Morrison had helped the Chinese government to negotiate a badly needed £10 million international loan, and he campaigned with some success in the British and international press on behalf of Yuan Shih K'ai and the Chinese government, predicting that the Republican government would rapidly transform China into a stable and prosperous regional power. One side-effect of this public campaign was that Morrison was written up in the British press as someone who could 'get things done in Peking' and Will was by no means the only person who wrote to him seeking his support for an initiative which would simultaneously benefit the Chinese government's reform programme and the proposer personally.

Will told Morrison that "*after my roving life in North China, England seems flat*" and sent him an album of photos showing the effects of deforestation in China. He was keen to use his arboricultural training in the interests of China, but despite an encouraging letter from the Chinese Foreign Minister, Dr Wu Ting-Fan (in pinyin, Ng Choy or Ng Achoy), he said that he was finding it difficult to get in touch with Chinese officials.† He was enclosing a paper‡ setting out his ideas to train working foresters in practical forestry and to set up a Chinese Forest School to educate supervisors to manage and to make better use of the forests: "*if at any time, the [Chinese] government should make a move in this direction and are desirous of foreign assistance, I should be most willing to offer my services.*"

Will's letter was lucid and well-argued (and far too closely aligned with what Han An was proposing to the Chinese bureaucracy in Peking for this to be a

---

* The cinema was barely in its infancy, and lectures such as this were a staple of entertainment in rural areas: it is likely that Will charged his audience for admission, generating a small income.
† This was not entirely true: Will was engaged in dialogue with Han An and possibly one or two other junior officials. But he hoped that Morrison would put him in touch with one or more of the senior officials or Ministers with whom he was on friendly terms. It is possible that Han An encouraged Will to write to Morrison because he hoped a high-level intervention from Morrison would break the log-jam of bureaucracy that was blocking the formation of a Forestry Department.
‡ Will's succinct and well-presented paper could be used in any modern management training course as an example of how to draft a 'pitch', and is reproduced in its entirety in Appendix B.

coincidence), but his timing was unfortunate. Faced with a National Assembly dominated by members of the Kuomintang party, Yuan had reacted by murdering those members who refused his bribes, thus crippling what might have been a forum for national reconciliation. He was also using a mixture of threats, murder and bribery to try – not very successfully – to bring the southern and western provinces under central government control. Three days before the date of Will's letter, seven provincial governors had declared themselves independent of Peking in what became known as the Second Revolution. The rebellion was crushed within a few weeks, but although Morrison replied to Will saying that he was *"really delighted with the beautiful photographs"*, and that *"there is nothing China requires more than a Forestry Department and there is nothing I would like better than to see you come out here for such work"* he concluded *"things are at present so unsettled that it is impossible to broach such a question."*[104]

In September 1913, Will wrote a 1,300-word front-page article on 'Plant-collecting in China' for the *Gardeners' Chronicle* in which he enthused about the *"haunting charm"* of the Peling mountain range.[105] He praised the beauty of the red-barked birch to be found over the Tibetan border, noted that *"many hundreds of seedlings are now established at the Coombe Wood nursery"*, and gave an account of the flowers of the region, including three different *Meconopsis*, six kinds of primula, and the tall honeysuckle with blue-black fruit *"from which excellent jelly can be made"*, concluding *"the above are but a few of the more important plants collected in this district"*. He recounted his successful hunt for *"Paeonia moutan in the wild state"* and described the grassland plateaux to the west of the Peling mountains, where *"showy herbaceous plants are scarce"* but *"a species of Senecio [with] pale yellow flowers [...] bids fair to become a useful subject"*. He concluded that

> *"the country is superb, of great geographical interest, and the climate bracing. Although, as has been said, the region was not so prolific of new varieties as more southern provinces it is hoped that those obtained, when worked up, will prove in due time of benefit to horticulture."*

Only a small number of the readers of the *Gardeners' Chronicle* possessed the background knowledge required to decode the sub-text behind the article, but this detailed justification by Will of the conduct and results of his 1911 collecting season would have been quite clear to Sir Harry Veitch and Ernest Wilson.

Remarkably, although the article gave a detailed account of his collecting in 1911, Will said not a word about the breakdown of law and order following the Xinhai Revolution, or the vicissitudes he had experienced. In referring to the local tribespeople, whom he noted were *"rougher than the average borderman"*, he merely stressed the absolute necessity for travellers, if they were to have any hope of being allowed to cross the mountain passes into Tibet, to become *personally* (my emphasis) friendly with their leaders and to gain their trust.

The *Gardeners' Chronicle* for the following week reported that *"owing to the expiry of the lease and impending retirement of Sir Harry J. Veitch"* the first auction of plants from Coombe Wood would be held at the nursery over the five days beginning Monday 13 October. The auctioneers noted that *"an important feature will be the offering of a quantity of the new and rare plants from China, some under seed numbers not yet being in commerce"*.

This auction (the first of three) of the Coombe Wood stock was attended by nurserymen and connoisseurs from all over Britain and Europe, who knew that the opportunity to purchase plant material of the quality and variety on offer would not occur again in their lifetimes. The catalogue listed over 2,700 lots, many of them consisting of dozens of plants and a total of 188 lots were identified as 'New Chinese plants'. Some may have been Wilson introductions, but most were plants collected in China by Will. Their dispersal amongst the commercial horticultural community without any indication that they had been collected by him dashed any lingering hopes Will may have cherished that as plants he had collected in China came to market so his reputation amongst potential future employers would be enhanced.*

Charles Hough procured many of the plants for the rock garden at White Craggs from the Craven nursery at Ingleborough, less than 40 miles from Clappersgate, which was owned by the writer and expert on alpine plants, Reginald Farrer. Farrer had travelled in the Far East and had made several plant-hunting trips to Italy and Switzerland. He had now been inspired by Wilson's and Will's expeditions to mount his own plant-hunting trip to China, primarily in search of alpines, to be funded by selling shares in the harvest of seeds and plants collected.

Farrer had first heard about Will's 1909–12 expedition in January 1911 when he wrote to Charles Sprague Sargent enquiring about Chinese alpines which Ernest Wilson might have collected. Sargent replied that he regretted he couldn't help because Wilson had not collected any alpines and had no plans to do so, but told Farrer that he and Harry Veitch had *"a man named Purdom"* collecting in northern China. He knew that Will had sent some interesting herbaceous and alpine plants, including primulas, to Harry Veitch and suggested that Farrer should get in touch with him.

There is nothing to suggest Farrer did contact Harry Veitch in 1911, but it would have been surprising if they had not talked about Will when they met at the RHS Primula Conference in London in April 1913. Veitch was exhibiting *Primula purdomii*, which won a first-class certificate and was much admired by Farrer, who also delivered a paper to the conference on 'Primula hybrids in nature'.[106]

The Director of the Royal Botanic Garden Edinburgh, Isaac Bayley Balfour, also attended the conference. He encouraged Farrer to pursue his idea of an expedition

---

* Two further auctions were held at Coombe Wood over the ten days commencing 2 February 1914 (5,000 lots, around one-fifth of them Chinese plants, none specifically identified as new) and the twelve days commencing 5 October 1914 (7,300 lots, 400 of which were identified as 'New Chinese plants').

CHAPTER 12

to China, where he was convinced there was much still to be discovered, but urged the need for an experienced companion, telling Farrer *"Purdom is the man ... if you can get him."*[*]

According to Charles Hough,[107] in the summer of 1913 Farrer asked him for an introduction to Will, and thereafter Farrer, Hough and Will *"spent an afternoon together at Kew"*. This must have been in May or June, before Farrer departed on a plant-hunting expedition in the Dolomites during July and August 1913. Farrer and Will discussed which areas of China were likely to be rich in alpine flora and the practicalities of plant-hunting there, but without formulating any definite plan. On 5 September 1913, after his return from Italy, Farrer wrote to Will that he had decided to go ahead with the expedition, seeking more detailed advice from Will about interesting locations and suggesting that he should join the expedition. Farrer could not, he explained, afford to pay Will a salary,[†] but would cover his expenses.

Will's reply was positive.[108] He identified an area of northern Kansu which *"might be worked to advantage"* and told Farrer that *"I am game for the proposal and should be delighted to accompany you and place any of my knowledge for the good of the undertaking"*. But he added that whilst in China he had endeavoured to interest influential Chinese officials in forming a Forestry Department, and he had continued to lobby in that sense since his return to Britain. It was *"just possible that I might be asked to help form such a department and get things under way."* If that came about, would it greatly upset Farrer if he quit the expedition after a twelvemonth? Farrer had no objection, and they agreed that Will would leave for China at the end of the year, to be followed by Farrer a month or so later.

It appears from Will's letter that his correspondence with Han An about returning to China to work in a yet-to-be-created Chinese Forestry Department was going well, although progress was slower than either of them would have liked. Although Will probably had sufficient savings to pay for a ticket out to China,[‡] he would have hesitated to incur the significant expense involved without a firm offer of employment. On the other hand, he would be able to promote the idea of a Forestry Department and his own bid for a post within it more effectively if he were on the spot. Farrer's offer to pay for Will to return to China and to bear

---

[*] Bayley Balfour never met Will, but as the Director of the Western centre of expertise on Chinese plants he closely followed the activities of plant-hunters in China, and was well-placed to offer advice about where to go and who might be available to accompany Farrer.
[†] Farrer had budgeted £1,000 a year for the expedition. He hoped to raise this by selling 'subscribers' ten per cent shares in the annual harvest of seeds and plants at £100 each but it seems that he found only six takers. He also contracted to write a series of articles for the *Gardeners' Chronicle*, which earned him c. £150 over the two years of the expedition, leaving him to make good a shortfall of a little over £300 a year.
[‡] Will's total salary for the entire duration of the 1909–12 expedition had amounted to a little under £700, with his day-to-day expenses in China reimbursed on top of that. It's likely that by the time of his return to Britain his savings were comfortably into three figures. But a one-way third-class train ticket to Peking would have cost Will £150 and it's doubtful whether the balance of his savings would have allowed him to maintain himself in China for more than two or three months.

his expenses for the duration of the expedition neatly solved this dilemma, and must have been a major factor in his agreeing to work without pay.

It is my belief that when Will set out on his return journey to Peking he intended to leave England for good. It was apparent from his enforced period of reflection in Westmorland that the rigid social hierarchy ruling his native land meant there was no realistic prospect of his achieving a measure of esteem and material comfort in Britain. During the previous 15 years he had worked conscientiously, and with some success, to pursue the avenues for self-improvement publicly advocated by British society – education, political engagement, and successful overseas exploration – and at the end of all his efforts he had quite literally found himself back where he had started. Like so many other young men (including his brother, Gilbert, who had emigrated to Canada in 1907), Will believed he could and would do better overseas.

Furthermore, Will felt great affection for China. He was convinced that he could, by helping to restore and develop the country's forests, improve the ecology of China and the quality of life of the Chinese people.* And he knew that in China he would not be subject to the ruthless pigeon-holing of the British caste system and would instead be judged by his actions and by the results achieved. When, at the turn of the year, he boarded the cross-Channel ferry, which constituted the first leg of the journey to Peking, it's unlikely that he looked back.

---

* The distinguished botanist William T. Stearn, writing in 1991, recorded that *"his [Will's] sister told me that he aimed to restore lost woodland to the Chinese and help them preserve their forest resources."*

Chapter 13
Back to China

Reginald Farrer, with whom Will spent 1914 and 1915 plant-hunting in China, is far better-known to contemporary students of horticultural history than Will, not least because of his books about the Farrer/Purdom Chinese expedition of 1914/15.

Farrer, who was born two months before Will, came from a prosperous Yorkshire family, many of whose (male) members were lawyers or clergymen. His father, James Anson Farrer, a dilettante barrister, was active in local government and an unsuccessful Liberal candidate in the 1895 General Election. When Farrer was nine years old his father inherited Ingleborough Hall, 37 miles south-west of Ambleside, the 4,000-odd acre Ingleborough Estate, and a further 10,000 acres of farmland and buildings, mostly in north-west Yorkshire.

Farrer was born with a hare lip and a cleft palate. As a child, he underwent a series of painful operations to enable him to eat normally and to speak intelligibly. He only achieved the latter at the age of 15, after the final surgery on his palate. Even then, his cousin Osbert Sitwell, who liked and admired him, described his voice as being *"as startling as the discordant cry of a jay or woodpecker [...] at its worst, like one of those early gramophones fitted with a tin trumpet."* [109] He largely grew up at Ingleborough, where he was educated at home by governesses and private tutors before reading Greats at Balliol College, Oxford.

From infancy, Farrer was a passionate botanist who became a self-taught expert – in his lifetime the leading British expert – on alpine plants,* and he made several trips to the Alps and the Dolomites in search of plants to bring back to Britain.

Farrer had a poor relationship with his father, who, after having agreed in 1903 to underwrite a trip by his eldest son to Japan (which included a side-trip of three weeks in Korea and China, where Farrer followed the standard tourist trail to visit the Forbidden City, the Great Wall and the Eastern Tombs), had been horrified by his extravagance and thereafter kept him on a tight financial leash. In 1908, Reginald travelled to Ceylon in order formally to convert to Buddhism, a decision which went down very badly with his parents. His love for his mother was immutable, but her subsequent efforts to bring him back into the fold of Evangelical Christianity severely tried his patience.

By 1913, Farrer's allowance had risen to £500 a year, still insufficient to allow him to set up his own home,† and he divided his time between Ingleborough Hall

---

* His books on rock gardening were best-sellers for decades. They are still appreciated for their lyrical descriptions of the plants mentioned, but botanists today are critical of their lack of scientific rigour.

† His books brought in a further £50 to £100 a year. He also ran from the Ingleborough Hall glasshouses the Craven nursery which sold alpine and Chinese plants: in 1916 (the only year for which figures are available) the business made a profit of just under £150.

CHAPTER 13

and the family house in London. Although he greatly resented his dependency on his parents and considered his allowance to be quite inadequate, it is doubtful whether they could have afforded to increase it. Against the background set out in Chapter 5, the Farrer family's decision in the mid-19th century to invest all their money in land and property was a serious error. Over the following hundred years the value of the land fell in real terms by around three-quarters, and the net profit realised by the Ingleborough estate during the period 1910–1915 was just £220.

There are other reasons for Farrer's deeply conflicted personality. He was homosexual* during an era when this was tolerated in private by some (not including members of his close family) but publicly universally execrated. Furthermore, 100 years ago the link between height and social rank was taken for granted, and Farrer was acutely sensitive about his short stature, to the point where he avoided being photographed standing next to anything which would enable observers to see that he stood around five foot three.[110]

At the same time, Farrer was a genuine polymath, a skilled self-taught watercolourist, especially of botanical specimens, the best-selling author of around 20 books and plays who possessed a truly remarkable command of the English language and an expert botanist.

The reverse of the medal is that Farrer was often vain and self-centred, consistently disinclined to share any public glory that might be going with anyone else, and overall determined to be seen by his peers as an aesthete and intellectually brilliant literary man-about-town with an impressive 'hinterland' in gardening and botanical exploration. He would not be happy that today he is remembered as an influential botanical writer and a successful plant-hunter whose novels and plays (several of which are much better than they are generally given credit for) are almost entirely forgotten.

But if there is one single fact that we must not lose sight of in considering Farrer's detailed accounts of the two years which he and Will spent plant-hunting in China, it is that when Farrer sat down at a typewriter it hardly ever failed to run away with him.† This is apparent from the verbose rococo prose of Farrer's books and some of his personal correspondence and also from the elegant but barbed *bons mots* which litter his writings and reflect the single least attractive aspect of his character, what his close friend and admirer Euan Cox described as *"a biting, and sometimes cruel, wit."*[111]

---

* After a good deal of hesitation, I have decided to use this word, rather than the contemporary 'gay'. One hundred years ago anyone identified as homosexual encountered active hostility from other members of society as well as the constant threat of criminal prosecution. It's hard to overstate how painful this was for the persons concerned. To be gay in Britain in the early 21st century is not the same experience as being homosexual 100 years ago.

† Although it is a comparison he would have simply hated, in that respect the short, plump, vain, Farrer resembles the short, plump, vain, Mr Toad in Kenneth Grahame's 1908 children's classic *The Wind in the Willows*. Toad was wholly unable to resist the temptation presented by the controls of a powerful motor-car: a typewriter keyboard was equally irresistible to Farrer.

In addition, Farrer was incapable of refraining from re-shaping an account of events in such a way as to achieve maximum dramatic effect and to present himself in the best possible light. For example, on the first page of the two-volume best-seller Farrer wrote about the 1914 season in north-west China, *On the Eaves of the World*, he recounts how his eye

> "*was caught by the unusual name of a lovely Primula* [Primula purdomi] *that dazzled the eye at the* [Primula] *Conference* [...] *Not a month passed before the same name caught my attention at a lunch-party* [...] *I besought a meeting* [...] *Immediately, I asked Purdom* [...] *whether he would be willing to share* [...] *a venture to the Tibetan March. In a couple of sentences the thing was settled.*"[112]

In short, Farrer starts his book with a classic introductory narrative trope of the hero who is unexpectedly presented with the opportunity to join a quest and at once, as heroes do, seizes the opportunity for adventure.

In reality the sequence of events was that the Primula Conference in April 1913 galvanised Farrer into doing something about turning his dream of plant-hunting in China into reality. He met Will, whose name he had first heard from Charles Sprague Sargent in January 1911 and who had been strongly recommended as a potential travelling companion by Isaac Bayley Balfour, in early summer and they discussed the practical issues involved in an expedition to western China. Farrer spent the next couple of months plant-hunting in Italy, thinking about just where in China he should go (initially he favoured Yunnan province) and persuading friends and acquaintances to subscribe to an expedition. Five months after the Conference he asked Will to accompany him to Kansu and the Tibetan border region. Will said he would think about it, and wrote to Farrer a day or two later accepting the invitation.

In truth, most of us are the heroes of our own legends and Farrer was not especially egregious in fine-tuning his own tale. Still, when reading his account of the two seasons he and Will spent collecting together, we should be aware that a degree of caution is called for in looking at the detail of his narrative.

Will travelled third class on the trans-Siberian Express to Harbin in northern China via Berlin and Moscow, arriving in Peking on 24 January 1914. He immediately re-engaged his former 'head man', a native of Shensi Province, referred to in Farrer's writings as "*the Mafu*"* and his brother. They set about putting in order the mule-boxes and other items Will had left with his friend Constable Pearson, who was in charge of security at the British Legation.

How and why Will retained ownership of this kit is obscure. In the normal course of events, one would have expected him to sell it before leaving Peking in 1912 and credit the expedition funds, and it is distinctly odd that neither Veitch nor Sargent suggested this. Whilst well-worn saddles and mule-boxes had little residual value,

---

* "*Mafu*" means 'head groom': it is regrettably typical of the travel literature of the period that this key Chinese staff member is never identified by name.

CHAPTER 13

other items, such as Will's firearms, would have realised a reasonable sum. Perhaps Will took the latter back to London and handed them over to Harry Veitch, which is what he may have done with the expensive Sanderson camera Veitch had provided him with. As for the motive behind Will's decision to store the equipment somewhere he could subsequently recover it, it's hard to think of any explanation other than that when he left Peking in 1912 he hoped to return.

Will also wrote a note to Morrison to say that he was back in Peking "*in advance of Mr Reginald Farrer, who follows me* [to Peking], *later in a trip out to S.W. Kansu*". Will hoped that whilst Farrer was studying high alpine flora he would be able to continue his research on the timber growing areas, work which he was in no doubt would, "*in time, bear good fruit.*" He asked Morrison if he and Farrer could consult Morrison's library and "*for a hand-shake*".

Farrer arrived in Peking in mid-February. Against the strong advice of the British Legation he and Will promptly applied for internal passports to go to western China.

The Legation's advice not to travel to western China was due to the state of violent anarchy prevailing in the region as a result of the activities of the bandit chief Bai Lang ('White Wolf'). Bai was an ex-soldier from Honan who had in 1911 joined a gang of robbers which was then co-opted by the southern Revolutionary leader Sun Yat Sen to join the 1913 rebellion. Bai's 'Army to Punish Yuan Shi K'ai' swelled to 50,000 strong and ranged freely across the western provinces, killing landlords and officials, burning tax records and sometimes redistributing wealth from the rich to the poor.

In early 1914 Bai over-reached himself by marching into Kansu, where the local Muslim warlords and their disciplined troops inflicted on him a series of stinging defeats. Bai's army responded by raping and killing Muslim civilians and burning Muslim villages, which made the conflict all the more bitter. In August 1914, Bai died and later in the year his army was crushed. In February 1914, however, this outcome was anything but certain, and Will and Farrer's refusal to deviate from their plan to collect in Shansi and Kansu was somewhere between courageous and reckless.

Farrer was sufficiently concerned by the risk he was taking that on 4 March he drew up his will, which was witnessed by Will and Constable Pearson.[113] He then spent most of the three-week wait for the passports to be issued shopping in the Peking curio shops, where he bought sable and ermine coats and a length of blue silk brocade embroidered with chrysanthemums, all of which he sent to Ingleborough. Will helped Morrison transplant some semi-mature trees into the garden of his 'country cottage' in the hills west of Peking and introduced Farrer to Morrison, who asked them both to write to him with news of the political and security situation along their route.[114]

In early March, Farrer, Will, their three Chinese staff, two ponies and several porters travelled 100 miles south-west to Pao-ting (now Baoding), where they spent the night before boarding a train to Mien-Chi (now Mianshi), in the southwest of Shansi Province.

Farrer and Will were invited to dine at the English Baptist mission station in Pao-ting, where they were surprised to meet Mary Gaunt, an intrepid Australian novelist and travel writer, who was hoping to cross China from east to west as far as Kashgar and then follow the Silk Road through Afghanistan, thence on to Moscow. Western diplomats in Peking had been unanimous in advising her not to attempt the trip, but she had been told by a friend who had travelled extensively in the region that *"Purdom knows a lot more about travelling in China than I do and if he says you can go, well, you might"*. Miss Gaunt records in her book *A Broken Journey* that *"Mr. Purdom was optimistic and declared that if I was prepared for discomfort and perhaps hardship, he thought I might go."* (She did go, but she was forced to turn back well short of Kashgar).

Will and Farrer – who had, on arrival at Mien-Chi added to their little caravan three mule-carts and a palanquin for Farrer's use – spent a month slogging through the muddy loess plains.* They were forced by illness to tarry a fortnight

Will and Farrer's carts bogged down in the mud on the way to X'ian, March 1914.
Royal Botanic Garden Edinburgh

---

* Loess is a fine wind-blown clay sediment which covers the north China plain, sometimes to a depth of hundreds of feet. Chinese roads at the time were for the most part no more than scrapes in the clay, iron-hard ruts in winter, deep bogs during the spring and autumn rains and choking dust in the summer.

CHAPTER 13

in the ancient city of Xi'an, but reached the mountains to the south of Lanchow, the provincial capital of Kansu, in early April. The local inns were *"sweet and wholesome"* and they botanised very successfully in the Nan-hor and Sha-tan-yu ranges. There, in the space of a few days, they found *Viburnum fragrans,* which Will had collected from a temple garden in Minchow and sent back to Harry Veitch four years before, *"in such situations that one could no longer doubt that here this most glorious of flowering shrubs is genuinely indigenous."**

Next came a fine hellebore (*Helleborus thibetanus*), described by Farrer as *"like a much-magnified Wood Anemone, with three or four dainty flowers of pearly white or pink spraying abroad from the one stem, each on a fine and dainty pedicle of its own, and with the fairy-like crimpled silk of Anemone nemorosa, instead of the rather fat and stolid consistency which usually prevails among the Hellebores. It has a grace, indeed, and charm of habit and colour and port which seem at present to my enthusiastic eye to set it apart in the race. But then I have not here for comparison the huger, stouter and showier hybrids in the group that are nowadays so freely raised and grown."*

This was rapidly followed by *Dipelta floribunda* – *"its bough bending beneath their burden of swollen-throated five-lipped bells of softest pearly white or faintest shell-pink, with a reticulation on the flower lip of what seemed like orange velvet"* – and by a yellow-flowered plant closely resembling a Daphne but which was later, rather briefly, identified as a new genus, *Farreria.*†

Best of all, on 18 April Farrer found a white form of *Paeonia Moutan*‡: *"tall and tender and straight, in two or three unbranching shoots, each one of which carried at the top, elegantly balancing, that single enormous blossom, waved and crimped into the boldest grace of line, of absolute pure white, with featherings of deepest maroon radiating at the base of the petals from the boss of golden fluff at the flower's heart. Above the sere and thorny scrub the snowy beauties poise and hover, and the breath of them went out upon the twilight as sweet as any rose."*

More discoveries followed, including the very fine *Buddleia alternifolia*§ – *"a most lovely plant, exactly like a very delicate weeping willow, but that the drooping sprays are set all along with little clusters of purple blossom, till the whole sweeping mass becomes a cascade of colour"* – and a smaller buddleia which Farrer christened *B. purdomii.*

Will photographed many of the above plants, and Farrer recorded the foliage and flowers in watercolours which were both scientifically accurate and beautiful.

---

* In 1966, the plant was re-named *V. farreri.* It was not called *V. purdomi* because in 1909 Will had found it in cultivation, and taxonomic convention is that plants are named for the person who first finds them in the wild or for some other characteristic, habitat, or locality.
† Later identified as *Daphne rosmarinifolia.*
‡ Now known as *P. rockii.*
§ Now known as *B. brachystachya.*

In addition, they prepared herbarium specimens and marked the plants so that seed could be collected in the autumn before the pair moved on to the semi-arid downs of the lower Min-shan range, briefly transformed by the spring rains into a mass of flowers.

But alpines were what they had come for, and alpines grow at high altitude, so they followed the line of what Farrer called the 'Blackwater' river (the Bailong Jiang) before turning west and crossing into Tibet, heading for the small village and monastery of Chago, a community with an unsavoury reputation and known to be hostile to foreigners but situated in a cleft between two limestone mountains, the ideal habitat for alpines.

The decision to turn off the main highway to Tibet towards Chago almost certainly saved their lives. Had they stayed on the main road they would have met the White Wolf and his army, burning and looting as they came, and would have been lucky to be taken prisoner for ransom. As it was, they arrived in Chago in early May and were not at all welcome.

Will in camp, probably on the Min-shan range, summer 1914.

Royal Botanic Garden Edinburgh

Some of the reasons for the near-universal hostility of Tibetans to foreigners at this time have already been touched on. Briefly, the Chinese government considered the Koko-nor* region of eastern Tibet to be part of China and 'Outer' (*i.e.* western) Tibet to be a tributary state. The Tibetan residents of the eastern region acknowledged Peking's authority to the extent of paying the annual tributes demanded by the Imperial Throne, whilst simultaneously paying tithes to the Buddhist lamas in the monasteries and temples, which housed up to one in five of the population.

On the ground, the authority of the lamas – known in the West, from the colour of the robes worn by the two senior classes of lamas, as the Yellow Church – far exceeded that of Chinese officials. But even the lamas exercised little control over the different groups of nomadic and semi-nomadic people, most of whom were not Buddhists but animists, who occupied large areas of otherwise empty countryside and who were more or less permanently in conflict with their neighbours.

---

* 'Koko-nor' could mean the Koko-nor lake ('the Blue Sea'), the town on the lakeside that grew up around the temple built by the Ch'ing administration in the early 18th century, or the region adjacent to the lake. Confusingly, the Peking government also referred to that region as the Qinhai district of Kansu province.

CHAPTER 13

The population of the Chinese territory adjacent to Tibet, which mainly comprised ethnic Tibetans, was only slightly more tolerant of foreigners than were the people on the Tibetan side of the border. Farrer and Will of course knew what had happened to John Weston Brooke in 1908.* They also knew what had befallen the plant-hunter George Forrest in the extreme north-west of Yunnan province during July 1905, when the local lamas murdered the French missionaries with whom he had been staying and made a determined but ultimately unsuccessful attempt to hunt Forrest down and kill him. They anticipated that the people of Chago would cordially dislike them, but they hoped that a well-founded fear of the dire punishment the Chinese government would inflict on the community if foreigners came to harm there would keep them safe. They also planned to curry favour by offering free access to their medicine chest and by paying in cash the asking price for food and accommodation.

Against this background, it is difficult to explain the circumstances of their entry into Chago. A mile away from the village and lamasery they came to a fork in the road. One way was smooth and direct but barred by hurdles of brushwood, the other route was circuitous, steep, and stony. Will and all the party save Farrer and one other took the latter route. Farrer made short work of the brushwood and rode along the better road before being reluctantly called to order by Will and retracing his steps to take the higher path. The best road was, it emerged, strictly taboo during this season in order to protect the crops from divine wrath in the form of hailstones. When Will and Farrer reached Chago, the inhabitants made their anger very clear.

After a hostile interrogation by the Prior of the lamasery, the expedition was allowed a verminous room. The following morning Will rose early and went for a walk. It was fortunate that he did not take his revolver, so that when he was faced with a hostile crowd shouting abuse and armed with smoking matchlocks† he had no alternative but to pretend indifference, pull his pipe out of his pocket, slowly fill and light it and walk back to the house whilst ostentatiously ignoring the mob. This proved exactly the right thing to do, as prudent or peaceable villagers urged the crowd not to harm someone who clearly presented no threat. But no sooner had Will crossed the threshold of the house than a group of lamas followed him to say, disingenuously, that whilst they had no problem with 'foreign devils', the villagers objected violently to their presence, and that for their own safety the party should withdraw. Will agreed, and the following day the little caravan made its way back down the mountain and into China.

It may be that the expedition would not have been able to remain in Chago even if Farrer had not flouted the taboo concerning the path. But the plain fact is that he had conspicuously failed the first test of his ability to live and work

---

* See page 96.
† *i.e.* ready to fire without the delay involved in lighting the fuse.

with the inhabitants of the high ground where the plants in which he was most interested grew. In Farrer's defence, he had spent most of his childhood in the village of Clapham, where his family owned 67 out of the 68 houses (the exception was the vicarage) and exercised near-feudal control over the villagers, nearly all of whom worked on the Ingleborough estate.* It was, arguably, inevitable that he should grow up firmly convinced that petty social constraints did not apply to him. But even Farrer understood that if Will had not defused the situation things would very likely have gone a great deal worse than they did. The incident caused a shift, perhaps not formally acknowledged, in their relationship, from chain-of-command to a partnership which would, in time, mature into genuine friendship.

Farrer waited in a small village, Ga-Hoba, just across the border whilst Will reconnoitred the Sha-tan (Farrer: Satanee) range a little to the west, returning to report that it was very similar to the mountains adjacent to Chago and the locals were friendly. They moved their camp to a dilapidated, almost abandoned, hamlet on the mountain-side called by Farrer 'the Miao of Satanee', from which they searched the mountain, finding a rose-coloured peony, a particularly beautiful rhododendron, several new primroses and a fine berberis. Despite unseasonal storms, they were about to shift to a camp in the alpine pastures near the summit of the mountain when the Mafu appeared to tell them that hail-storms had destroyed the crops at Chago, clearly a sign of divine wrath at the blasphemous activities of the foreign devils, and that, as paraphrased by Farrer "*the whole rage of the district was directed against ourselves*". They made the painful decision to abandon the Satanee range, barricaded themselves into the house against an expected night attack which never came, and in the morning made their way by a circuitous route to the little walled town of Siku (now Zhouqu).

Siku nestled at the foot of the Leigu Shan mountain, which in his writings Farrer called Thundercrown and, further to the north, the Min-shan range, where Will and Farrer hoped at last to be able to settle down to systematic botanising. The two governors (one civil, one military) of the town had been ordered by Peking to watch out for the expedition: they were happy to see them safe and well but were initially reluctant to allow them to venture outside the city walls. In June, however, the tide of war turned decisively against the White Wolf and the plant-hunters were able to spend two weeks on Thundercrown, where they found a rich harvest of anemones, primulas and poppies, including the elegant little harebell poppy, *Meconopsis quintuplinervia*. On 6 July they left Siku by the north gate and made their way on empty roads and through the charred ruins of villages

---

* If this sounds too strong, consider the young estate worker at Ingleborough who was told that he would not, as expected, be spending the next week working and living in a bothy at the other end of the estate and who was sacked on the spot for 'cheek' when he said this was a shame because his mother had already made up a bag of food for him, or the worker who was told by Mrs Farrer that if he missed Sunday church again he would lose his job, and with it his home. The other side of the coin was that the estate provided the village with piped water, electricity, and community facilities which included a school and reading-room.

CHAPTER 13

Two of Will and Farrer's Chinese team, 'Ma the Mohammedan' (left) and 'Lay-Go the donkey man' (right). Lay-Go accompanied Will on the hunt for Dipelta and Buddleia seeds and would have done the talking if they had been challenged by any locals.

Royal Botanic Garden Edinburgh

devastated by the Wolf's army to Choni, from where they botanised very successfully for the rest of the summer on the Min-shan range.

On 4 August 1914, innocently unaware that Britain and Germany were at war, Farrer and Will braved fine but persistent rain to collect seeds and herbarium specimens of the blue/purple 'Lonely Poppy', *Meconopsis psilonomma*.[115] Or did they? Farrer says that they did, but there is strong circumstantial evidence that Will found the plant a few days earlier and that the herbarium specimens were collected two or three weeks later.[116] It seems that Farrer, writing in 1917, could not resist using the beauty of the flower to contrast the last tranquil prelapsarian days of peace with the horrors of World War 1.

For the first nine months of 1914 there was no mail service of any kind in Western China, and no news of the outside world beyond an occasional telegram shared by a local missionary. Farrer and Will had no inkling of the developing European crisis, and when, in late August, they met William Christie in Minchow and he told them of the war between France and Germany,* it came as a complete surprise. A few days later, their mail from Peking caught up with them and they learned that Britain was also at war. Like many others, some of whom should have known better,† they believed it would 'all be over by Christmas'.

In September, the emphasis changed from collecting plants and herbarium specimens to collecting seeds, a back-breakingly laborious business which Farrer described with feeling in the *Gardeners' Chronicle* and which required re-tracing their entire season's itinerary. They had found in the Min-shan range duplicates

---

* Germany declared war on France on 3 August 1914 and Britain declared war on Germany the following morning. But several foreigners record that the first reports of the conflict to reach China were of war between France, Germany and Russia, with no mention of Britain.

† For example, one junior interpreter at the British Legation in Peking wrote to the Foreign Office asking for permission to break his contract in order to return to London and join up, and was refused on the grounds that by the time he had worked out his three-month notice and made his way home (taking at least a further month, now that the rail route via Berlin was closed) the conflict would have ended.

Will in the disguise he wore to collect *Dipelta floribunda* and *Buddleia alternifolia* seeds in the hills above Chago, September 1914. © British Library Board

CHAPTER 13

of almost all the plants discovered in the neighbourhood of Chago, including *Dipelta floribunda* and *Buddleia alternifolia,* but had collected very little seed of the former and none from the latter. Both are exceptionally attractive plants, and Will and Farrer were determined to obtain seed from the specimens they had marked on the hills above Chago. Clearly, this had to be done without the knowledge of the villagers. According to Farrer, he and Will decided to disguise themselves as Chinese labourers to collect seeds and for what would be a long day's journey there and back to the inn at Ga-Hoba. Will then regretfully pointed out that Farrer's gold front teeth – which had been much admired by locals during the summer – were incompatible with any such disguise, and he would therefore have to make the trip alone. (Quite apart from the teeth, Will may not have been truly sorry about this because of Farrer's limited ability to understand, let alone speak, Chinese.)

Will borrowed a set of clothes from their head mule-driver, grimed his face with burnt cork, and set off with the donkey-man, a local who would do the talking if they met anyone. The trip was entirely successful – according to Farrer the only Chagolese encountered on the hills fled in terror when confronted by "*a ruffian of daunting stature and murderous mien and most dour and dreadful countenance*" – and Will returned with a sack of Dipelta seed-pods* and a fat packet of Buddleia seeds. The next day, before they left for Siku, Will, wearing his disguise, used the clockwork timer of his camera to take a series of four self-portraits.†

---

* After sifting, however, the whole sackful yielded "*the merest pinch*" of tiny seeds.
† One of these is the photograph referred to in the Introduction. It was printed in the *Gardeners' Chronicle* under the headline, "*Polymetis* [cunning] *Odysseus*" and was for many years the only photograph of Will in the public domain.

BACK TO CHINA

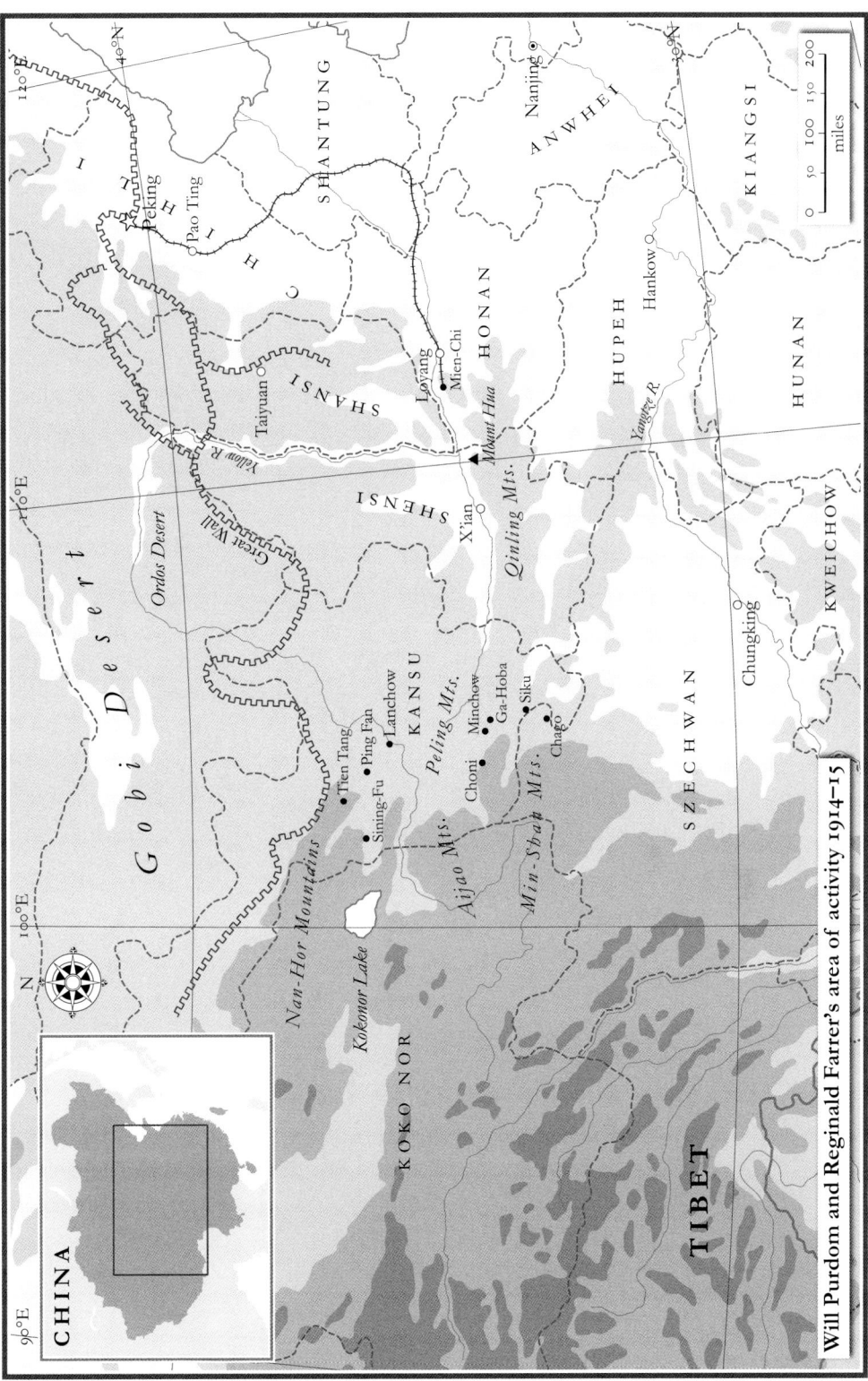

Chapter 14
Winter 1914/15

Will and Farrer reached Siku on 22 October, weary but content after a satisfying season's collecting. On arrival they learned that, by a remarkable coincidence, there was another Western botanist in town. Farrer was not pleased: he subscribed whole-heartedly to the prevailing etiquette observed by European plant-hunters whereby the collector who was the first to reach a discrete area was entitled to exclude his rivals until such time as he saw fit to move on, and during plant-hunting expeditions to the Alps and Apennines he had used his sharp tongue to enforce this rule.

Farrer was somewhat mollified when he learned that the person in question was the US Department of Agriculture collector, Frank Meyer, who had travelled to Ambleside to meet Will just under two years before. Meyer was in search of plants which would be useful to American farmers and market-gardeners and was not competing for ornamentals with Farrer and Will. He intended to stay in Siku for only a few days before going on his way across Kansu and Szechuan into Tibet, where he hoped to find hardy peach trees and almond bushes. When, on 23 October, Will and Farrer called on Meyer and his Dutch assistant, Johannes de Leuw, the meeting was reasonably cordial, although Meyer was preoccupied with problems with his staff. Four of his muleteers had already refused to travel to Tibet and he was concerned that his interpreter might do the same. The following day, his interpreter, Tien Chi-nian, whom Meyer had selected in Peking from 19 candidates, to whom he had carefully explained the precise itinerary he intended to follow and whom he had paid handsomely in advance, did indeed announce that he had heard *"all sorts of horrible stories"* about the terrifying denizens of Tibet and would go no further.[117]

Here it is necessary to say a little about the languages of China 100 years ago. Educated Chinese men were expected to be able to read and write Classical Chinese, the thousand-year-old language of the Tang dynasty, and this was the written language of the court and senior administrators. But over the course of the millennium, the pronunciation of the language in northern and southern China had diverged to the point where an educated gentleman from Shanghai could only use it to communicate with his counterpart from Peking by an exchange of written notes, a practice known as a 'brush conversation'. Mandarin, the day-to-day language of educated people in Peking, and whose relationship to classical Chinese could be compared to that between modern English and the language of Chaucer, was widely used as 'the language of the officials' for day-to-day administration, but it was completely unintelligible to folk in, say, Canton or Szechuan.

CHAPTER 14

The different tongues spoken in Canton, Fuzhou and Shanghai at least had their roots in classical Chinese, in more or less the way the European Romance languages are rooted in Latin. But the language of parts of Shensi and of Kansu was a linguistic outlier, related to Burmese and spoken by almost no one not native to those provinces. Not the least of the services Will had rendered to the expedition was to re-engage his former headman, a native of Shensi, and his younger brother. They had both lived for some time in Peking and were able to communicate with Will and as necessary with regional officials in Mandarin and with the natives of Kansu in the local language. Few Chinese people were able to do this, and the sub-set of that number willing to endure the discomfort and danger of a journey across the far west of China into Tibet was even smaller. The defection of Meyer's interpreter, who would be almost impossible to replace before rapidly approaching winter made travel in the region out of the question, was a devastating blow to his plans.

Meyer could not compel Tien to accompany him, but was determined to make him refund the wages he had been paid and to secure damages for breach of contract. He went to the town magistrate to argue his case, but Tien had the better of the initial exchange. Will and Farrer were out of town on a last ascent of Thundercrown, but on their return they accompanied Meyer to the magistrate and explained the situation in more detail than Meyer had been able to.* The magistrate forbade Tien to leave the town pending a decision on damages but, according to Meyer, Tien bribed the constabulary and made good his escape.

So much for the facts. What happened next was that Farrer took up his pen and drafted an elegant and witty account of events, which he promptly sent off to the *Gardeners' Chronicle*. He discoursed smugly about the unwisdom of bringing "*up country*" from Peking such a "*fine gentleman*" of an interpreter, and reported that when Meyer had reached Siku and told Tien the proposed itinerary, Tien had flatly refused to go another step. "*An altercation followed and a rapid descent of the stairs by the interpreter*". Meyer, according to Farrer, had arrogantly refused to learn Chinese and was left "*perfectly helpless without a servant and with no prospect of getting another*". Fortunately, he and Will were at hand to intercede on Meyer's behalf with the Governor, find him a servant† and "*send him on his way rejoicing.*"

By the time Meyer found out about Farrer's account, which did not appear in print until July 1915, it was far too late to complain. The version which appeared in 1917 in *On the Eaves of the World* incorporated further embellishments, including a Chinese man who tried to act as peacemaker between Meyer and Tien and was punched on the nose by Meyer for his trouble. It's beautifully written and has become the single most frequently cited anecdote about Meyer's character and his time in China.

---

* Will did the talking: at this time Farrer's grasp of Mandarin was almost non-existent. Even after his vocabulary had improved, Farrer's speech impediment must have made it very difficult for him to articulate the five different tones which were an essential element of the spoken language.

† The young man in question was an opium addict whom Meyer sacked after a week.

Although it had been Will who helped Meyer make his case to the authorities, Farrer was thoroughly familiar with the true facts of the matter. However he chose deliberately to distort them in order to present himself as a prudent and knowledgeable gentleman-traveller who had helped the lumpen, not to say crass, Meyer out of the mess his loutish behaviour had got him into.

Will does not seem to have had any great liking for Meyer, but he recognised the injustice done. The thought must also have occurred to him that if Farrer would insouciantly trash Meyer's reputation solely to present himself to better advantage then it might be wise to be careful about sharing with Farrer any personal anecdotes which might tempt him to do the same thing to Will.

Farrer's room in the house he and Will rented in Lanchow during the winter of 1914/15: it is apparent that Farrer enthusiastically collected objets d'art as well as plants.

Royal Botanic Garden Edinburgh

Will and Farrer agreed that Will would go alone to the Ajiao mountains (Farrer: Arjeri) at the end of the Min-shan range, 75 miles to the north-west of Siku to collect seeds. They would then meet in ten days' time at the little village called by Farrer 'Bow-u-Go'. Farrer spent a few more days gathering seed on Thundercrown before leaving Siku on 30 October. He arrived at Bow-u-Go a few days later, and waited there until Will arrived with a good haul of seeds, whereupon they made their way to Lanchow, where they proposed to spend the winter.

The road to Lanchow took Will and Farrer past the holy mountain of Lien Hwa S'an,

> "*a shattered limestone mass of eleven thousand feet, aspiring in countless little spires and pinnacles into the naked blue, each one of which is crowned with a little shrine, only to be attained by rock ladders and perilous chains of iron driven into the precipices.*"

Farrer recorded that Will made a detour to visit "*a grubby little old Daoist nun*" who had nursed him back to health when, on his first expedition, he lost his footing and fell and stunned himself, but

> "*the Wolves had had their way of the little old nun, and when Purdom reached her cell in its grassy glade he found only a burnt and blackened circle on the sward and a scatter of charred bones and a carbonised skull.*"[118]

This is the only reference to Will having injured himself during the 1909–12 expedition – none of the letters he wrote at the time mention such an incident –

CHAPTER 14

Will and Farrer in the Lanchow garden of the Viceroy of Koko-nor, Zhang Bing Hua (centre), late 1914/early 1915. This is the only known photograph of Will and Farrer together. © British Library Board

and Farrer's account does ignore the obvious question of what Will's Chinese staff did immediately after he fell off the iron ladder. One suspects that whilst the story may have some basis in fact, it lost nothing in Farrer's re-telling.

Will and Farrer arrived at Lanchow in mid-November. They rented a large and comfortable house for the winter which they shared for ten weeks with a young American, Ronald Gilbert, who was overwintering on his way north.* Gilbert had first come to China in 1912 as a 21-year-old patent-medicine salesman and had made enough money that he was able to stay on, learn Chinese to a good standard, and re-invent himself as a journalist. Will and Farrer also made the acquaintance of the other members of the local expatriate community, including a postmaster from Yorkshire, Mr Boyer, and his wife, and several staff from the mission station headed by William Christie. Farrer made formal calls on the military governor, General Wu, and the Viceroy of Koko-nor, Zhang Bing Hua, both of whom possessed fine gardens they were pleased to show the visitors and which Will photographed.

Farrer had kept a daily diary during the previous six months, which he now used to write the first draft of his 200,000-word account of the first year of the expedition, *On the Eaves of the World*, which was published in 1917. Although Farrer never acknowledged Will's contribution to the text, the book draws heavily on Will's knowledge of Chinese language and customs. It also includes descriptions of the countryside which rely on the experience of Will's 1909–12 expedition. Some of these passages read very much as if they were written by Will rather than by Farrer. For example, the description of northern Kansu and the Koko-nor lake (which Farrer never saw, but which Will had visited in 1911), expressed concerns about the impact on local rainfall of deforestation and the benefits of reforestation, issues about which Will felt strongly but in which Farrer is not known to have been interested:

> *"It is certain that at one time the now-ruined country must have been much better watered and wooded in general. Nowadays, thanks to the destruction of every twig for ages past, Northern Kansu is as desert as the Sahara, bleak and torn and hopeless [...] the verdant flats of Lanchow are encircled as far as eye can see in wrinkled dust-yellow ranges of stark lifelessness [...] Even the Koko-Nor [lake] is yearly shrinking, and the ruin of a world approaches visibly unless some large and serious steps are taken promptly, such as have so miraculously saved and reclaimed the New Province, thanks to a Governor who planted so widely and well that within two generations that once imperilled land has perceptibly started changing for the better."* [119]

Amongst their letters were a quantity of back copies of the *Gardeners' Chronicle*, including that for 3 October, which included a letter from Will's friend Charles

---

* In May 1914, a story had appeared in the *Los Angeles Times* saying that Gilbert was on his way to Liangchowfu (now Xining) *"in the hope of uncovering a treasure believed to have been hidden centuries ago by the bodyguard of Confucius under the* [city] *wall"*. It is impossible to know whether Gilbert believed this nonsense or whether he cynically hoped that the story would make his 'copy' more attractive to foreign newspapers. In any event, there was no treasure and the story never really got off the ground.

CHAPTER 14

Hough commenting on the first of the series of Farrer's articles about the expedition. Dr Hough was indignant that Farrer had so far not so much as mentioned Will's participation in the expedition and told the editor that

> "*It will interest many of your readers to know that Mr. W. Purdom took an active part in organising the expedition [...] As you will be aware, Mr. Purdom has previously explored much of the ground already covered and any measure of success [...] that attends the labours of this exploring party will be largely due to the experience previously gained by him. I feel sure Mr. Farrer will be the first to acknowledge this.*"[120]

Farrer fired off an immediate riposte to the editor (it was published six weeks later, the time taken by the letter to reach London) protesting that although Dr Hough no doubt meant well,[*] his letter was ill-judged and plain wrong. Dr Hough was, he said, almost openly accusing him of "*burking*[†] *the credit of my companion, Mr Purdom*". He went on to say

> "*I call the expedition mine because it so is – planned and arranged by me and joined by Mr. Purdom on my invitation. These facts I had judged to be common knowledge [...] I do not see any reason for making my published articles a puffing advertisement of my companion or myself. My own work may not deserve such: Mr Purdom's stands above need of it*".

Farrer concluded by contesting the statement that much of the ground had already been covered by Purdom: this was "*quite untrue and calculated seriously to depreciate the value of our exploration.*"[121] Nonetheless, Dr Hough and, presumably, Will must have been pleased to see that whereas there was nothing in the first eight of Farrer's 32 articles in the *Chronicle* to suggest that Farrer was not collecting in China alone, thereafter he started to include references to Will and his activities.

The claim by Farrer that he had planned and arranged the expedition must have caused Will some surprise, but he fully understood Farrer's concluding point. The expedition had been funded so far by the sale of shares in the first year's harvest, the seeds and tubers he and Farrer were busily sorting and packing. The two mule-loads they had brought to Lanchow would enable them to send a very satisfactory quantity of plant material to George Redman, the manager of the Craven Nursery, to be forwarded to subscribers or 'grown on' in the greenhouses of Ingleborough Hall. Importantly, this material included seeds of seven primroses, five androsaces,[‡] and five poppies, all of them new to Britain, of *Buddleia alternifolia,* likewise new to

---

[*] Had Dr Hough not been the leader of a syndicate which had bought a share in the expedition, one suspects that this nod to courtesy might have been omitted.

[†] The reference is to the 'resurrection man' William Burke, hanged in Edinburgh in 1829 for the murder by suffocation of several people whose bodies he then sold for dissection by medical students. The verb was briefly current in Scotland in the sense of to suppress or to keep quiet. Farrer's revival of what by 1914 was a long-forgotten idiom brilliantly distracted readers from the unconvincing argument advanced in his letter, namely that Will's fame was such that any mention of him in Farrer's account of the expedition was superfluous.

[‡] English name 'rock jasmine'.

Britain, and seed of a large number of plants of which there were very few specimens in Britain, sometimes just one or two in a botanical garden.

This was what the expedition subscribers expected and what they were paying for. They did not merely want a reasonable quantity of seeds; they wanted, and had been promised by Farrer, seeds of plants which none of their friends and acquaintances who were not also shareholders in the expedition could hope to acquire. The seeds were much prized by subscribers for precisely that reason. Years later, one of Dr Hough's young daughters described sitting at the dining-room table with her father and sisters dividing up the contents of the packets, "*hardly daring to breathe in case we blew the seeds away.*"[122]

The weekly *Gardeners' Chronicle* was the gardening publication of choice for professional gardeners and horticulturalists, and for the middle- and upper-class owners of country estates who wanted to keep up with developments in botanical science and new plant introductions,[123] including the people buying the seeds and plants which Farrer and Will were collecting and sending back to Britain. The suggestion that what Farrer was sending to subscribers was, in the main, plant material which had already been introduced to Britain by Will could all too easily lead potential purchasers of shares in the 1915 season to conclude they might just as well buy plants from those nurserymen who had successfully bid in the 1913 and 1914 Coombe Wood auctions for Chinese plants. That would have effectively cut off funding for the expedition. In the event, though, this concern was overtaken by the crushing effect of the war on British gardening.

Farrer and Will were now receiving the London *Times* fairly regularly, though six weeks late, and exchanging letters with friends at home. It was increasingly apparent from the former that the war would be protracted and from the latter that it was having a devastating effect on gardening and horticulture in Britain. Professional gardeners had been just as keen as other young men to join the 1914 rush of volunteers for the armed forces* and competition for workers from factories making military equipment or munitions made it impossible to replace them. Coal and coke to heat greenhouses were unobtainable. Horticulturalists and nurserymen were being told by the Board of Agriculture that if they did not at once dispose of their stock, if necessary by ploughing it in, and start growing vegetables, then the authorities would take over their land and glasshouses and do it for them. The final nail in the coffin of the British horticultural trade was that spending money on exotic plants in time of war had overnight come to be seen as frivolous, if not downright unpatriotic.

Partly for the above reasons, at least two of the shareholders in the 1914 season did not now wish to buy shares for the second year. One, Robert Woodward, an

---

* For example, by the end of September 1914, eight weeks after war had been declared, 56 out of 88 male staff (there were also 22 female staff members) at the Royal Botanic Garden Edinburgh had enlisted. By the end of the war, the total had risen to 73, of whom 20 were killed on active service.

enthusiastic arboriculturist who had in 1912 bought from Veitch & Co seeds and plants collected by Will, had joined up and was serving in France.* Another, Walter Fenwick, had died in July 1914 and although George Fenwick-Owen, the big-game hunter befriended by Will in Choni and Minchow in 1912, had taken over his cousin's [124] 1914 share, he did not consider himself obliged to subscribe for 1915.†

Farrer's hopes of balancing the books during 1915 now rested on his dialogue with the Liverpudlian cotton-broker and nurseryman, Arthur K Bulley. Bulley had declined to buy a share in the expedition when Farrer approached him in 1913, but after reading the mouth-watering accounts in the *Gardeners' Chronicle* of new and lovely plant discoveries and shortly before the outbreak of hostilities he wrote to Farrer offering to purchase a substantial part of the 1914 and the future 1915 harvests. Farrer wrote back offering half his 40% share of the 1914 collection for £400. The letter crossed with one from his mother telling him that the Craven Nursery would have to close.‡ Meanwhile, the New and Rare Plants Department of Bee's nurseries, owned by Bulley, was selling fewer and fewer seeds and plants. After a series of offers and counter-offers which left Farrer feeling he had been badly taken advantage of, he accepted Bulley's offer of £600 for 95% of his share of the 1914 harvest and the totality of his share of the 1915 collection.

Farrer had also written a long letter to Professor Bayley Balfour, the Regius Keeper of the Royal Botanic Garden Edinburgh, promising him some interesting seeds and botanical specimens. He hoped many would be new to Western botany and urged Balfour to name some of the best and most attractive of them *purdomia* or *purdomii*. Writing in confidence, he told Balfour that

> "between the lines of his story, he [Will] *had bad luck on his Veitch-Sargent trips, appalling and almost killing hardships and undergone alone, and in the end not always a very kindly or adequate treatment of the plants he had toiled out his life to get.*"

Farrer set out at length the "*almost embarrassing*" extent to which he was in Will's debt, saying "*It is he, after all, who alone makes the expedition possible*" and deploring that "*he is now among the many thousands beggared by this awful war!*" He hoped that "*if we ever get back*" Balfour would meet Will and "*see whether some nook might not be found for him.*"

The pen-picture Farrer, an acute observer, drew for Balfour of Will's character after a year spent working closely with him, for long periods in the complete absence of other English-speaking company, deserves to be quoted in its entirety:

---

* In March 1915, George Redman, Farrer's manager at the Craven nursery, sent Woodward's share of the 1914 harvest to his home at Arley Castle, Bewley. There is a copy of the accompanying list of seeds and plants in the Kew archives, but Woodward never saw Farrer's notes or the seeds: he was killed in action on 9 May 1915. He was posthumously awarded the Military Cross.
† Fenwick-Owen, who also won an MC, was wounded twice but survived the war.
‡ This did not happen: the business continued to trade until at least 1921.

"He is, as you know, a propagator and cultivator second to none but above all a man of character and pleasantness impossible to over-praise, but of a fiery loyalty (a devotion to everyone else's interests that deprives him of the specially supple neck and knee that some 'great' men demand, and all great institutions prefer.) Put him, however, under a keen and sympathetic superior, in a place where he was allowed to vote Liberal and where all the crossing-sweepers had a fair living wage and I am sure any garden would bless the day that admitted him to reign over it. He is far too big for a private place or I would strain every nerve to have him at Ingleborough: as it is, I shall always rely on his help there."

The only surprise in the above is the reference to Will's supposed Liberal sympathies. There is no reason to suppose that Will's socialist beliefs had changed since his days at Kew. It is possible that Farrer was deliberately 'not frightening the horses' in what was for all practical purposes a character reference, but it seems more likely that Will had maintained a degree of reticence about his politics and had allowed Farrer to believe that he shared his Liberal views. The terms in which Farrer recounted to Balfour Will's travails with Veitch and Sargent also suggest that Will had given him no more than a bare outline of events. Will had seen in October how Farrer had been unable to resist turning Meyer's difficulties in Siku into a gripping but disobliging tale, and this may have prompted a degree of caution on his part.

It appears that Will did not know about Farrer's letter to Bayley Balfour, which may have been just as well. Farrer, motivated by pure kindness, was asking Bayley Balfour if he knew any person or institution, who would, in the light of Will's botanical and horticultural ability, overlook his obdurate refusal to 'know his place' and display appropriate humility when dealing with his social superiors and engage him *"under a keen and sympathetic superior"* as *"a propagator and cultivator"*. This was emphatically *not* the kind of work Will aspired to. Indeed, he had spent most of the previous decade working to break into the kind of *"superior"* management post which it evidently did not occur to Farrer to suggest could be filled by someone of Will's social class.

Although Will would no doubt have recognised that Farrer was genuinely trying to be helpful, if he had known the terms of the letter his reaction to Farrer's special pleading on his behalf would surely have been on the lines of 'Keep your patronage; what I want is a fair chance to apply for any work I am competent to undertake'. In any case, although Bayley Balfour's reply promised that if he heard of a suitable appointment he would certainly think of Will, he told Farrer that he saw no chance of anything suitable coming up in the Civil Service in the foreseeable future, nor of any post falling vacant at Edinburgh during the remainder of his term of office. It wasn't a cheerful prelude to the 1915 season.

Chapter 15
The 1915 season and return to Peking

The question why, even after it became clear that the war would last years, not months, neither Will nor Farrer packed their bags and returned home to enlist is inescapable. In Farrer's case, at least part of the answer is simple: he didn't meet the minimum height requirement for the armed forces.* In addition, his medical history made him unfit for military service. His deep personal commitment to Buddhism meant he was profoundly opposed to violence. Finally, he did not want to cut short the expedition which fulfilled his dream of plant-hunting in China and he sincerely believed the scientific importance of what he was doing was more important than any contribution he could make to the war effort.

Will felt guilty that he was not 'pulling his weight' in the war, but he was unwilling to abandon the expedition, which would have collapsed if he had withdrawn. Farrer self-evidently lacked the language and organisational skills to go it alone. Will also shared Farrer's belief in the scientific value of the expedition and wrote to an old school friend in Ambleside:

"Alas, alas, I'm too far away at present to be of service to my country but I trust in years to come my little I give to the progress of the world may last in the memory of mankind, if it be only through a fading flower." [125]

Will's view that his contribution to botanical knowledge might, in the long run, be seen as more significant than anything he could have achieved by military service is one which may surprise, even shock, those of us who grew up in 1960s Britain, where popular culture embraces almost without question a heroic vision of both World Wars, but it would not have struck many of his friends back home in that light.

The initial rush to enlist in 1914 briefly obscured deep divisions in British society, which were then almost immediately thrown into sharp relief by the war. In 1915, the 'Derby scheme', an attempt by the British government to establish a viable alternative to conscription by persuading men of military age to 'attest', *i.e.* undertake to join the armed forces on being invited to do so, was a conspicuous failure: more than one in three single men and one in two married men refused to attest. Conscription was imposed in 1916, but many more men avoided the draft than the post-1918 narrative of shoulder-to-shoulder engagement by the British people in the 'war to end wars' admits. Neither of Will's brothers ever wore khaki, and at least one of his close friends from his student gardener days at Kew,

---
* In September 1914 the minimum height requirement was raised to five foot six inches, although in late 1914 'Bantam' units were formed for volunteers over five foot tall.

CHAPTER 15

William Johns, was amongst the thousands of men who went to live in Ireland, where conscription was not applied, and waited out the war there.

Insofar as it is possible to generalise about the motives of the large but under-researched cohort of young men who deliberately avoided military service, their refusal to serve was not solely due to a desire to avoid a one-in-three chance of death or life-changing injury.[*] In many cases it was also a deliberate snub to a government which they had no say in electing, since only men who owned a dwelling-house with a rateable value over £10 p.a. or who paid more than £10 p.a. rent could vote. Farrer, for example, was entitled to vote in General Elections,[†] but Will was one of the 40 per cent of the male population which was not. Will was, moreover, a committed socialist who wanted to change the British political and social system rather than to fight to defend the status quo. Finally, the shabby way he had been treated by the botanical and horticultural establishment after he returned from China in 1912 may still have rankled.

Will and Farrer spent early January finishing the sorting, cleaning and packing of the 1914 harvest before sending 138 large envelopes of seeds and 15 boxes of plants to Farrer's manager at the Craven Nursery, George Redman, who would send all but one of the subscribers their shares. This substantial consignment testifies to the skill of both men and their sheer hard work during the 1914 season and is all the more impressive when one recalls that it does not include George Fenwick-Owen's ten per cent share, which he asked should be sent direct from China, and Will's personal share, another five or ten per cent.

In mid-January, Will rode south through the snow to Choni in search of a pair of antique copper cooking pots which Farrer coveted. Will may also have felt he needed a break from the company of Farrer and their house-guest Ronald Gilbert, especially the latter. Gilbert's attitude to Chinese people was condescending at best:[‡] together with his heavy drinking and frequent recourse to the local brothels, this made him an uncongenial companion.[126] Will returned triumphant in late February with the pots and a quantity of other treasures.

The weather turned, the frozen Yellow River thawed, and on 29 March Will and Farrer left Lanchow for the little walled town of Sining (now Xining) east of the Koko-nor lake and very close to the Tibetan border. Once there, Will told Farrer that he would now ride out to explore the nearby mountain ranges in search of a suitable base for the next season, with only the Mafu for company.

Three weeks later, haggard and exhausted, Will returned to report that after finding the lamas at one attractive spot relentlessly hostile he had sought advice

---

[*] 6.2 million British men served in the armed forces in World War 1. 745,000 died on active service. In 1921, 1.2 million were in receipt of disability pensions.

[†] In fact, Farrer could vote twice, once for an MP representing the constituency in which he lived and once for an MP representing the University of Oxford.

[‡] Gilbert's 1927 book *What's wrong with China* is a deeply unpleasant pseudo-scientific justification of imperialism by reference to alleged Chinese idleness, incompetence and deceitfulness.

Will and Farrer's mule-train on the ridge above Wolvesden House.
Royal Botanic Garden Edinburgh

Wolvesden House, beautifully situated half-way up 'Wolf stone valley'.
Royal Botanic Garden Edinburgh

from William Christie, and approached the monks of the Tien Tang monastery around 80 miles to the north-east.* The official passports from Peking and the letters of introduction provided by the Viceroy of Koko-nor had persuaded the Abbot to give his blessing to botanising by the expedition over the whole of what Farrer called the Gadjur range, unexplored and apparently promising terrain, and he had identified and rented in its entirety a suitable inn on the eastern slope of the mountains for six months.

Will needed to rest for a few days to recover from the rigours of the journey, but they left Sining on 1 May. A few days later, after crossing two high snow-covered passes, they reached the inn which would be their base for the summer. Will told Farrer that the Chinese name of the valley in which the building sat was 'Wolf Stone Valley', and Farrer promptly christened the place 'Wolvesden House'.

For quite different reasons, neither Will nor Farrer had ever had a home of their own and they both enjoyed the process of transforming a scruffy set of mud and timber buildings, whose design resembled nothing so much as five or six dilapidated shipping containers arranged around a central court, into a

---

* Will's sister Nell, writing in 1977, said that Will had been advised by a local missionary, "*Mr Christiansen*", where to collect in 1915. I can find no missionary of that name listed in the comprehensive records of the different missionary societies active in northern China 100 years ago, and I believe that "*Christiansen*" was a slip of the pen for "*Christie*".

The yard at Wolvesden House.
Royal Botanic Garden Edinburgh

Looking down on Wolvesden House from the ridge.
Royal Botanic Garden Edinburgh

snug dwelling. Their staff, not all of whom welcomed the hard labour involved, shovelled decades of mule droppings from the rooms and the courtyard, which they cobbled with stones from the nearby stream, and made openings in the mud walls for windows, glazed with spoilt glass photographic plates washed clear. One of the two metal stoves which they had brought with them (at 11,000 feet, Farrer recorded *"there was never a day, even in midmost August, when one could even think of dispensing with the stoves."*[127]) was installed in the 'drawing room' and one in the kitchen. After a fortnight's hard work, they were in possession of a warm living-room furnished with bookshelves and tables for writing up notes and sorting seeds, two bedrooms complete with *k'ang* sleeping-platforms, a spare bedroom, a kitchen, staff quarters, a darkroom for Will to develop photographs, and an area of the yard ready for 'heeling-in' cuttings and live plants.

This is as good a place as any in this narrative to consider whether the relationship between Will and Reginald Farrer may have included a homoerotic element. One of Farrer's biographers has suggested, on the basis of the *"ripening campness"* of several references to Will in *On the Eaves of the World*, that this might perhaps have been the case.[128] The difficulty with this argument is that richly camp was Farrer's default literary style. His publishers had on at least one occasion made him remove words and phrases they considered to be unseemly from one of his books and some of his personal correspondence would be considered 'over the top' today.[129] The brief passages in which Farrer remarks on Will's good looks and describes his dressing up in furs or antique armour must be seen in this context.

Will was a handsome man, but there is not a scrap of evidence that Farrer was attracted to him. Nor is there anything known about this or any other period of Will's life to suggest that he would have reciprocated any such feelings. Indeed, it's arguable that what little circumstantial evidence there is points the other way, since during the course of the expedition Will regularly absented himself from Farrer's company for weeks at a time. It is said that Farrer appreciated these periods of absence because, as a non-smoker, he did not much enjoy sharing what was usually rather cramped living space with such a heavy smoker as Will. None of this is consistent with an intimate relationship.

Whilst the work on Wolvesden House was going forward, Will and Farrer received an emissary from the Tien Tang Abbey formally making them free of the mountains, to which they returned thanks and a ritual scarf of friendship. They arranged a weekly delivery and collection of mail from the post office at Sining-Fu and regular deliveries of milk and coal. When on 18 May they set off on their first field trip, their spirits were high and they were optimistic about what they might discover during their second season.

Spring arrived in fits and starts, and rather than living under canvas whilst they botanised on the other side of the Tatung (now Datong) river, Will and Farrer stayed at the Tien Tang Ssu monastery, called by Farrer *"the Halls of Heaven"* and the larger Abbey of Chebson Ssu, where they were welcomed as honoured guests and made very comfortable. Will initially discouraged Farrer from obliging the Prior and sub-Prior at Chebson with whisky-and-soda, despite their broad hints about *"foreign wine"*, fearing that the Abbot would react badly if his foreign guests were discovered to have corrupted monks who had taken a vow to abstain from alcohol. He relented when the lamas explained that medicinal alcohol was permitted, and that they both had upset stomachs, and shared evening drinks became a fixture at the end of the day's collecting.

Personal comfort aside, the valley of the Tatung river and the mountains on each side of it were disappointing. The mountains were granite, not the limestone preferred by most alpine plants, and compared to the previous year the pickings were decidedly thin. They did find a new primula and a new and beautiful *Isopyrum* (sometimes called 'false rue anemone'), *I. farreri* and, at the end of the season, a lovely new pale blue gentian which Farrer named *Gentiana farreri* and which he considered the single best plant discovery of the expedition.*

The beauty of the scenery went some way to offset the scarcity of new plants: in late spring and summer the valleys were carpeted with flowers, including clematis, iris, poppies, gentians, primulas, saxifrage and delphiniums,† from all of which they collected seed. On 5 July they were both stunned into silence when they climbed over the crest of a small valley above Wolvesden House to find a grassy pasture

---

* Farrer was right: this beautiful, vigorous plant, fully hardy in Britain, is much prized by British gardeners.
† By an odd series of coincidences, nearly all the flowers they saw in 1912 were either blue or purple.

CHAPTER 15

three or four miles across covered as far as the eye could see with shimmering, pale blue, harebell poppies. Recounting the incident later, Farrer described how Will turned to him and half-whispered, *"Doesn't it make your guts ache?"* concluding *"It was exactly right, no other words could have nailed the truth so absolutely"*.

In early September Will returned to Sining-fu to collect the mules they needed to remove their belongings from Wolvesden House. There he was offered several fine fur coats which he suspected had been looted during the 1912 Revolution. Will, a lover of fine clothes, bought the lot, and he and Farrer enjoyed trying them on when Will returned to Wolvesden House.

Farrer and Will then stripped out of the building the additions they had made, packed up their belongings and left Wolvesden on 13 September for Ping-Fan, far to the south on the upper Luo river. They had planned to hire a boat there to take them to Lanchow but the river was too low, so they rented a cart and followed the riverside track instead.

At Lanchow they discovered that in October General Wu had been replaced as military governor by General Ma Ch'i and the friendly Viceroy of Koko-nor had been arrested for accepting bribes, although Will suspected the real reason for his incarceration was that as a Manchu his loyalty to the government of Yuan Shih K'ai was considered to be doubtful.

At about this time, Farrer received a letter from Arthur Bulley saying that in Britain the demand for seeds and ornamental plants had fallen away completely. Bee's Nursery was only surviving because he was cross-subsidising it from his cotton-broking business. Bulley could not now pay £600, as agreed, for most of the 1914 and all the 1915 harvest, but only half that sum. This was a crushing blow, but there was absolutely nothing to be done about it.*

Will and Farrer spent several days at Lanchow crating up their specimens, the furs and antiques and most of their other possessions, all of which they arranged to be taken by cart to the railway bridgehead at Mien-Chi for onward shipping to Peking. Rather than return to Peking by the route that they had taken on the outward journey, they decided to make their way three hundred miles south-east to Lo-yang (modern Luoyang) at the confluence of the Luo and Yellow rivers in Shansi Province, thence by boat to Chungking and on to Peking. This would allow them to botanise along the Kansu/Szechuan border as well as to see the spectacular Yangtze Gorges. First, though, Will would make a final sweep through the Ajiao mountains and the area around Choni in search of seeds of which the season's harvest had been poor. He would then rejoin Farrer in Lo-Yang.

Farrer reached Lo-yang after a three-week trek through the mountains and waited several days for Will, who eventually arrived weary but laden with seeds. The river was still too low for boats, so they followed the carters' highway south

---

* It says a good deal for Farrer's character that he recognised Bulley was recounting the simple truth, and that on his return to England he and Bulley resumed friendly relations.

through increasingly hot and humid climes to Chungking and boarded a steamer to Hankow. There they caught a train to Peking. On 6 December they checked into the 'Grand Hôtel des Wagons-Lits' and squared up to three months' mail. The next day Will called on Morrison who recorded that he was *"full of beans, but bored stiff with Reginald Farrer"*.

The news from home was bleak, a catalogue of friends and acquaintances killed in action, including one of Farrer's closest friends from Balliol, Alfred Gathorne-Hardy. David Prain's son had been killed at Ypres and Isaac Bayley Balfour's son had perished at Gallipoli. Balfour nonetheless wrote stoically to Farrer, praising the quality of the seeds and herbarium specimens sent by the expedition, which Farrer attributed to *"the very high standard aimed at by Purdom and diligently imposed upon his Chinese 'boys' in 1911"*.[130]

There was one piece of good news: as Farrer told Balfour, it appeared that Will might secure a permanent job in China as Forestry Adviser to

"His Imperial Majesty Yuan the First* [...] such work is just exactly what he [Will] would most take pleasure in [...] he would be incessantly up and down in the wilds, arranging for plantations and cuttings. His eyes will always be towards the Alpine hills and I guess he'll do a good deal of studying of forest conditions up the Thibetan [sic] border if only to cast a sidelong squint at the Primulas!"

---

* On 13 December 1915, the same day that Farrer wrote to Balfour, Yuan Shih K'ai proclaimed himself the first Hongxian Emperor of China. Farrer's ironic tone when referring to the Imperial title displayed sound judgement: on 22 March 1916, widespread opposition forced Yuan to quit the Dragon Throne after a 'reign' of just 83 days.

Chapter 16
Forestry Adviser to the Chinese Government

Many of the men who were appointed to senior government posts after the proclamation of the Chinese Republic were what we would today call technocrats. They were strongly committed to modernising Chinese society and governance so that China could take its rightful place in the community of nations and were convinced that this could only be achieved if China followed the example of Japan in acquiring and exploiting the scientific and industrial knowledge of the West. This was certainly the view taken by Chang Chien (in pinyin, Zhang Jian) who in September 1913 was appointed Minister of Agriculture and Commerce. Chang enjoyed a formidable reputation as a highly successful industrialist who had used the profits from his businesses to fund numerous social and cultural institutions, including an agricultural training college, in his native Nantung. Immediately after taking the reins at the Ministry he had greatly reduced his department's running costs by merging into one the administrations of what had been two separate government departments. President* Yuan Shih K'ai considered him to be one of the most capable members of his Cabinet.

In mid-December 1915, Chang persuaded Yuan that a Chinese Forestry Service applying the scientific forestry pioneered in Germany and in British India could simultaneously reduce China's dependence on imports of timber, help to reduce flooding in the countryside, and showcase the benefits the unified Republic would bring to the Chinese people. Ernest Morrison, meanwhile, lobbied the Minister of Finance, Chou Hsüeh-Hsi (in pinyin, Zhou Xuexi) to support the scheme, including the proposal to employ Will as one of the Directors of the new service.

The Chinese Forestry Service was formally established in January 1916, by which time Chang had resigned in protest at Yuan's proclamation of himself as Emperor: his replacement was another moderniser Chow Tzu-Chi (in pinyin, Zhou Ziqi). On 10 January Will signed a three-year contract with the Chinese government as Divisional Director responsible for forestry training schools, tree nurseries and anti-flooding projects.[131] His salary was 600 Chinese dollars a month,[132] equivalent to c. £800 p.a., although due to the government's chronic shortage of funds, salaries were often paid substantially in arrears. The Director General of the service was the Deputy Minister of Agriculture, but day-to-day management was in the hands of two Co-Directors, the first of whom was Han An,

---

* I am following the practice of both Chinese and Western historians in continuing to refer to Yuan Shih K'ai as President rather than Emperor.

CHAPTER 16

the Secretary to the Minister, and the second William Sherfesee, an American who was the former Director of Forestry for the Philippine Islands.

The mandate of the Service was to work to make timber more readily available and cheaper, to reduce flooding and silting-up of rivers, to protect and manage existing forests, to educate *"all classes of people"* about the need for reforestation, to form a corps of Chinese forestry experts and to encourage private businesses, including by providing advice and support, to engage in similar activities.

Will must have been deeply happy to have at long last broken through the glass ceiling and be appointed to a senior post where he could pursue objectives to which he was personally strongly committed, and he promptly called on Morrison to thank him for his support.[133] Morrison took the opportunity to ask whether Will would come to his country cottage to look at the fruit trees he had helped transplant in 1914, and Will promised to do so.

Both Will and Morrison believed that Sherfesee was not competent to discharge the post of Forestry Service Co-Director to a good standard. On 26 February 1916 when Will encountered the Morrisons on a weekend walk outside Peking, Morrison noted that

"Already he [Will] has been able to form an opinion regarding the new Adonis on Forestry, the American Sherfesee. No more unsuitable appointment could be imagined. S. is a keen assertive Southerner, purely an office man who knows nothing about work in the field, nothing about trees, nothing about botany. He has set himself to transfer to Peking the whole office system of the Philippine Forest Bureau. Really glad must be the U.S. govt. to have got rid of such a man [...] He has begun badly is badly fitted for his work and will accomplish little or nothing."

A rather scrappy notebook of Will's from this time survives,[134] and records that on 11 March he travelled to Xi'an to inspect a possible site for a tree nursery, a project which the city governor enthusiastically supported.

Farrer remained in Peking until 18 April 1916, when he set off on the difficult and dangerous journey back to Britain, by train to St Petersburg then through neutral Finland and Sweden to Norway, and across the mine-ridden North Sea to Hull.

On the day before his departure Farrer wrote a farewell letter to Ernest Morrison[135], with whom, and with Mrs Morrison, he had enjoyed friendly relations over the previous four months, to the point where he had been invited to dine at their home on Christmas Eve. (Morrison's diary records waspishly that *"Farrer came half an hour late but was amusing, if unpleasant to look at and hear speak"*.)

Farrer expressed his thanks for the way in which Morrison had *"enriched my stay"* and went on to commend Will to his care. He particularly hoped that Morrison would protect him from those expatriates who were taking advantage of Will's good nature *"to get a first-class gardener for nothing"* and cited a recent instance when *"some peculiarly underbred woman"* had told Will that since he owed his job in the Forestry Department to her, he should expect to work for her "*at her pleasure*". Farrer claimed that he did not know the identity of the woman concerned, but it was clear from his detailed account that he was referring to Mrs Morrison. He closed his letter by expressing regret that he (Farrer) had only been able to secure for Purdom a post as Divisional Director of the Forestry Service: if only Morrison had sponsored Will he would surely have been appointed to "*the principal post*" instead of the "*frantically pro-German*" American, Sherfesee.

Farrer no doubt believed that he had delivered a corrective reprimand which was both elegant and effective, but his letter enraged Morrison. He immediately wrote to Farrer in Yorkshire:

> "*Yesterday morning I received your letter dated April 17th. Rarely have I received a more offensive communication with its cheap sarcasm and insulting insinuation, made still more odious by your action in arranging that it should reach me only after your departure*".[136]

Morrison flatly denied that he had expected Will to do any work for him for nothing: the only person who had exploited Will was Farrer, who had for two years profited from his expertise and hard work without paying him a penny.

Morrison went on to tell Farrer that he had heard about Farrer's disgraceful behaviour from Will after his return to Peking in December, when Morrison (*not* Farrer) had helped him secure well-paid employment with the Chinese government. On Sunday 9 April, Will had indeed come for afternoon tea at Morrison's country cottage, and spent just 20 minutes walking around the garden (in a particularly nasty aside, Morrison said "*I did not tip him: perhaps there is his grievance*"). As for Sherfesee, he was doing his work to a high standard and it was quite wrong of Will to traduce his chief as he had. Morrison was writing to Will to demand an explanation for his gross misrepresentation of the facts to Farrer.

Morrison's letter to Will of the same date[137] was even more intemperate. He listed grievances about Farrer's putative misbehaviour during 1914 and 1915 which Will had allegedly shared with Morrison on 18 December: Farrer was selfish, incredibly mean, rude to English missionaries, drank to excess and participated in "*drunken orgies*" in Lanchow and elsewhere. Farrer had made Will "*ashamed of being an Englishman*" and Will had told Morrison that "*for no money on earth would you travel again in the Interior with such a man*". Morrison had never repeated these stories and had treated Farrer with perfect friendliness, as he had Will. Will had repaid him by falsely telling Farrer that because Morrison had helped Will secure a well-paid job, he expected Will to work for him for nothing. Finally, Morrison warned Will that he had made a bad beginning to his work in the Forestry Department by "*bringing serious charges against the head of your department*".

Will replied briefly to Morrison denying that he had misrepresented him, or to stand behind anything Farrer had said or written. He denied all the points made

CHAPTER 16

by Morrison regarding their conversation or regarding his departmental chief and concluded by saying that

> "The Chinese have been good enough to warn me about being cleared out of the job so those interested in that little deal will be saved quite a lot of trouble. I will be much more use at home just now".

Farrer's reply to Morrison, dated 14 May,[138] did not reach Peking until 11 June. Farrer did not address the substance of Morrison's complaint but chose instead to criticise his prose style in patronising terms. He began by expressing his confidence that by now Morrison would have regretted writing in the terms he had employed, which were no doubt attributable to "*the cares and nerve-strain of your unfortunate position.*"* Morrison's letter "*asks for the charity of oblivion.*" Farrer knew Purdom too well to believe "*the colouring you try to put on our relations*" and "*you may be sure that whatever the source of one's* [Morrison's] *information, it is never Purdom!*" He concluded by hoping that Morrison would "*return to a kinder temper*" and added a postscript saying that he was sending Morrison's letter to Will "[so] *that he may appreciate its skill.*"

Over a century after the events in question, there is little point in trying to disentangle from the above precisely what was said to whom and when, let alone what may or may not have gone on in Lanchow during the winter of 1914. Morrison's diary for 18 December 1915 does, however, confirm that he and Will walked to his country cottage and Will told him "*funny stories*" about Farrer, including that Farrer drank to excess, and had been rude to missionaries.[139] And in February Will had told Morrison that Sherfesee didn't know much about trees or botany, when it appears Morrison had concurred.

In his letters to Farrer and to Will, Morrison also confirms that Will rather reluctantly visited the cottage on 9 April and looked round the garden with him. The following day was Will's 36th birthday and one can easily imagine that, in the course of celebrating with Farrer, Will may casually have mentioned that the previous day he had gone to look at Morrison's garden – he hadn't really wanted to, but Mrs Morrison had insisted and made some comment about his owing it to her husband, who had helped get Will his job. Within a day or two Will would have forgotten all about the conversation, and literally the very last thing he expected, let alone wanted, to happen was that Farrer would turn his laconic remarks into an account of ill-treatment so flagrant that Farrer was duty-bound to intervene to make all right.

The consequences of these bad-tempered exchanges are less hard to elucidate than whatever truth lay behind them. Farrer did Will no favours whatsoever

---

* Farrer was referring to the generally held belief in the Peking expat community that in 1912 Morrison had quit a position of great influence (*Times* correspondent in Peking) for a post as Special Adviser to the Chinese government, which was far better paid but where he had no real power because his advice was often not sought or ignored. We know from Morrison's diary that he worried that he had made a mistake by resigning from the *Times*: Farrer's condescension, and the tone of considerate sympathy in which it was delivered, rubbed salt in the wound.

by thoroughly antagonising on his behalf a leading member of the expatriate community in Peking, someone whom Will had been cultivating for at least six years, and who had used his influence with the Chinese government to support Will's bid for a senior post in the Forestry Service. Morrison now evicted Will from the corner of the field next to his cottage which Will had been using to cultivate plants from his share of the 1915 harvest, and until the end of his life his occasional references to Will in letters to third parties were consistently disparaging.

One immediate effect of the row with Morrison was that Will broke off communications with Farrer, ignoring his letters and even telegrams. We may suppose that Will was somewhat embarrassed that Farrer should have found out, at least in outline, what Will had told Morrison about his conduct during 1914 and 1915, and angry and upset that Farrer should have taken it upon himself to make a drama out of a trivial incident and then, with no concern for the consequences of his actions, to elbow his way to centre stage in his role as Will's self-elected guardian angel. Although after this exchange Morrison does not seem actively to have used his influence to hinder Will's career in China, he was clearly not going to be helpful when Will's contract expired in 1919. The loss of Will's proving-ground was a nuisance, as was the awkwardness he experienced when he found himself attending the same social events as Morrison and Mrs Morrison. Will knew Farrer well enough to know that he hadn't intended all this to happen, but that was hardly an excuse.

The conflict between Will and Sherfesee which Will mentioned in his final letter to Morrison was serious: Will's letter implies that Sherfesee was manoeuvring behind the scenes to get him sacked; we may also speculate that Will was unwilling to take orders from someone he considered manifestly incompetent. The row caused Will to submit his resignation, but the Ministry refused to accept it and the senior management structure of the Service was changed into a *de facto* triumvirate of equals comprising Han, Sherfesee and Will. All three men dealt directly with Ministers and all of them were referred to as 'Forestry Advisers to the Chinese Government'. Will was for the most part busy 'in the field', and Sherfesee was exclusively concerned with administration and finance in Peking, and this arrangement seems to have worked smoothly.

On 19 and 21 April Will attended meetings at the Ministry of Agriculture headquarters in the buildings of the Temple of Heaven concerning the site for tree nurseries, including one near Peking (Will thought the gardens of the Old Summer Palace were very suitable, "*plenty of water, and soil good*"). Later in the year, he travelled around northern and central China to establish nurseries and hire staff to run them, going at least as far north as Mukden (now Shenyang) and as far south as Hunan Province.

The death of Yuan Shih K'ai on 6 June 1916 and the indictment, a few days later, of Chow Tzu-chi for high treason to the Republic, brought an abrupt halt to the work of the Forestry Service. Work eventually resumed under a new

CHAPTER 16

Minister of Agriculture, Ku Chung-Hsiu (in pinyin, Gu Zhongxiu), but he owed his appointment to his long-standing opposition to Yuan Shih K'ai and did not share his predecessor's enthusiasm for scientific forestry. Indeed, one contemporary press report hints that the Minister had links with timber industry interests which saw the nascent Forestry Service as a threat, and that he deliberately weakened it for his personal advantage.[140]

Things looked up somewhat in August when Will and Han An went on a tour of Chihli province (modern Zhili), the province in which Peking stands, to draw up a reforestation scheme to reduce the flooding which every year killed thousands, often tens of thousands, of people. Up to a million trees were planted in the Western Hills in order to reduce run-off but this was an isolated initiative: Han, Will and Sherfesee remained significantly under-employed.

Sherfesee used his free time to write a series of long articles on the importance of forestry and reforestation in China, China's Forest Law, and the organisation of the Forestry Service. Will, although he had shown by his 1912 and 1913 articles in the *Gardeners' Chronicle* that he was perfectly capable of writing interesting and readable prose, felt that to date the Forestry Service had not achieved enough to justify writing about it, and chafed at the lack of work.

Will's persistent reluctance to go into print – for example, he did not use his period of virtual unemployment between mid-1912 and late 1913 to write a book about the 1909–12 expedition – testifies to his reluctance to puff up his work in the public arena. In this respect, Will was almost unique amongst professional plant-hunters of the period, all but one of whom published more or less self-serving accounts of their expeditions.* One consequence of this reticence was that when the Forestry Service was in the doldrums Will had very little to occupy his time. He found it hard to cope with inactivity, and especially with periods when he was unable to go out into the field and get his hands dirty by handling plants and plant material, and it seems that in the course of 1916 he became seriously depressed.

The news from home was not calculated to make Will feel any better. 1916 was a dreadful year for Britain. The attempt to open a new front at Gallipoli was finally abandoned in January, after British and Australian forces had lost 42,000 men killed and 100,000 wounded. The Easter Rising in Dublin was suppressed with a cack-handed brutality which made the existing tensions between different communities on the island of Ireland even worse. On 1 June, the Royal Navy defeated Germany's High Seas Fleet at Jutland Bank, but Britain lost 15 ships and over 6,000 men in the battle. German losses were 11 ships and 2,500 men and the Germans, at least initially, also claimed victory. On 5 June, the cruiser taking Lord Kitchener to an Allied summit in Russia was sunk in the North Sea by a German mine. Kitchener drowned with the rest of the ship's company and the death of this iconic figure was a shattering

---

* Roland Cooper (1890–1967), who collected very successfully in India, Bhutan and Burma between 1913 and 1922, also never published a memoir. As a result, he is even more thoroughly forgotten than Will.

Tea party hosted by Will for the household of his Manchu landlord, 1917.   Purdom family papers

blow to national morale. The Somme offensive, launched on 1 July, was abandoned in October. It had cost 125,000 British soldiers their lives and a further 300,000 had been wounded. At home, the cost of living had almost doubled in two years. Popular support for the war was being eroded by despair that it seemed as if it would never end.

Characteristically, after writing his riposte to Morrison's angry letter, Farrer had completely forgotten the whole business. He had been surprised and upset when during the summer and autumn of 1916 he wrote to Will from Yorkshire and from London but received no reply. He asked friends at the British Legation in Peking for news of Will, and on 5 November he wrote to Will's mother on the pretext that he wanted to send Will some albums of photographs of their expedition and needed to know whether they should go to the Legation in Peking by diplomatic bag or would Will prefer them to go to Ambleside?

> "Of course, I should prefer to send [the albums] *straight to him, but neither by hook nor crook, by letter or cable, can I elicit the smallest answer or sign of life from him, so that of his actual whereabouts I know nothing, nor of how he is getting on: you would greatly oblige me, indeed, if you would supply me with what news there is of him. I know he is no correspondent, but I do confess I very much want to know that he is well and happy wherever he is.*"[141]

The reply from Jane Purdom is lost, but in December Will did write to Farrer, when he described himself as *"fed up with things in general."*[142] He regretted that

CHAPTER 16

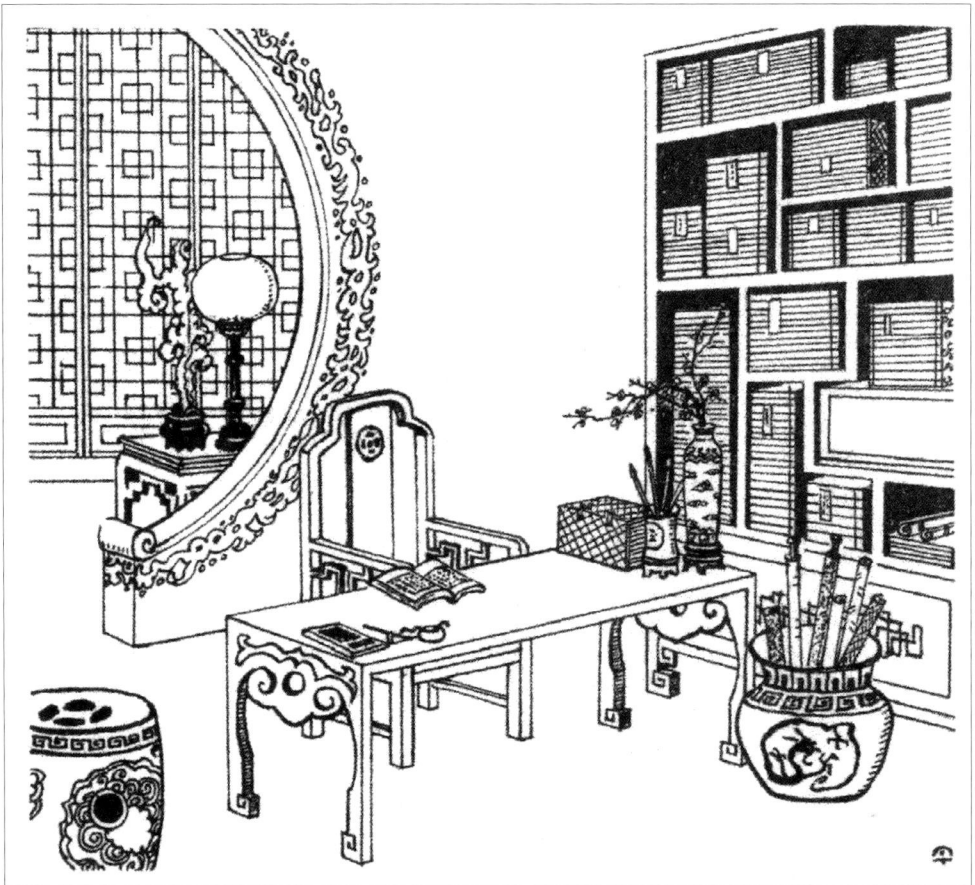

Sketch of Will's study in Peking, 1917. Nell Purdom/Sweet Stone Bridge

he could not be off again to collect more treasures: "*Peking is about as bad as being in a prison.*"

Will complained to Farrer that, in addition to not having enough to do, he was being paid in notes worth about half their face value. He had no contact with the Legation, for fear of being seen as "*on the make*" and had sub-let the house he had rented earlier in the year and moved to live in an inn with a Chinese friend.* He was very pleased to hear about the welfare of the plants they had collected together and by those novelties which the taxonomists at the Royal Botanic Garden Edinburgh had named *purdomii*, but closed by saying that he had been sorry to read in the press that Farrer's friend and contemporary at Balliol, Raymond Asquith had "*gone west: it is a bloody war.*"† He added a postscript, a

---

* On receiving the letter, Farrer underlined "*Chinese*" and "*inn*". For an Englishman to move in with a Chinese friend in Peking was definitely not 'the done thing', any more than was having his home in a Chinese inn.

† Raymond Asquith, son of Herbert Asquith (Prime Minister 1908–16), a brilliant Classics scholar who was a contemporary of Farrer's at Balliol, was killed on the Somme on 15 September 1916. It is unlikely that Asquith and Will ever met, but Will would have heard Farrer speak about him.

prose poem recalling the happy days spent collecting in the wilds of Kansu and Tibet and concluding bleakly, *"No more, by God."**

The plain fact that he wasn't achieving very much in China made it increasingly hard for Will to justify to himself his decision not to return home and join up. His avoidance of the members of the diplomatic community in Peking reflected his dislike of the frivolity of diplomatic receptions and cocktail-party chatter, but it also enabled him to avoid pointed questions about why he had not 'answered his country's call'. Since Will fell squarely within the scope of the second Military Service Act of 1916, which conscripted into the armed forces all British men aged between 18 and 41, it was even possible that the Consul might invite him to collect his call-up papers, as would happen to George Forrest when he was plant-collecting in Burma in 1918. (Luckily for Forrest, the war ended before the paperwork reached the Consul at Tengyueh.)[143]

Will's next letter to Farrer was written on 11 April 1917, the day after his 37th birthday, and is not entirely coherent. Mixed in with some confused gossip about mutual acquaintances in Peking are references to how tiring life is in Peking *"after many seasons in the GLORIOUS WILDS"*, to the liking rabbits have for Farrer (who might have complained in a letter to Will about damage to his plants in Yorkshire?), and an exhortation to Farrer to *"watch and pray, for the time is not yet"*. Will concluded by apologising for a *"wandering, disconnected note"*, which he attributed to a post-birthday hangover.

Will's letters home were also causing some members of his family serious concern about his mental equilibrium, to the point where his youngest sister Nell† travelled to Peking to 'look after him'. Nell was living in London with Will's elder sister, Margaret, 'Peg' to her siblings and parents. The two sisters were both teaching in a north London girls' school of which a year or two later Margaret, who was 11 years older than Nell, became the headmistress. It appears that Margaret funded Nell's ticket to China from Will's London bank account, which she still managed on his behalf.

If her later life is anything to go by, Nell was an adventurous young woman who would not have taken much persuading to quit her job in London and travel to Peking. She may also have had at the back of her mind the thought that, given the drive to educate girls in China, it would not be difficult for her to find an interesting teaching position there.

No correspondence from this time survives, but Nell arrived in China on the eve of the Moon Festival, 30 September 1917, after a month-long journey via Marseille and the Red Sea. The shipping companies had adopted this route to minimise the likelihood of encounters with warships of the Triple Alliance or with mines laid by them, which would have been much harder to avoid on the pre-war route from

---
* The complete text is at Appendix C.
† The third of Will's four sisters (born 1889) was christened 'Nellie', but liked to be called, and always signed herself, Nell.

CHAPTER 16

Southampton across the Bay of Biscay to Gibraltar and the Mediterranean, Even so, the petite (barely five foot) bespectacled 28-year-old Nell displayed real grit in undertaking this long journey in time of war and both she and Will must have been relieved when she arrived safely in Peking.

According to the rather saccharine memoir she published privately 50 years later, Nell found Will in fine form, living in gentlemanly elegance in the wing of a Manchu palace which he was renting in the West City. The palace was undoubtedly elegant, and by the end of 1917 Will may genuinely have been in better health than he had been in the spring and summer of that year: Nell did write to Farrer that Will was *"much brighter"* since she came to Peking, and had put on weight. But the reality of Will's situation was probably better captured by Farrer in a letter he wrote to his mother in February 1918 telling her that

> *"I had a letter the other day from Nell Purdom, who has at last gone out to take charge of her derelict* brother. She confirms me in my fear that a life of solitude in Peking has not agreed with my poor Bill's nerves or health."*[144]

It is not easy to extract hard facts from Nell's episodic and rather superficial account of her stay in Peking between 1917 and 1918, but the absence of any reference, direct or indirect, to the war, not even to the truly spectacular firework display organised by the Chinese government to celebrate the Armistice, suggests that this was a painful subject, deliberately avoided. Also, when Nell wrote to Farrer on 20 April 1918 she commented that *"He [Will] has a goodly circle of Chinese friends who are infinitely more charitable than most of our own race and foreign parts"*.

With all due caution, we may infer from Nell's comment that Will had been taken to task by some members of the British and perhaps the French, Belgian or American communities in Peking for 'shirking his patriotic duty'.[†] The likelihood of criticism on these lines had hardened to near-certainty after November 1916, when the former chief engineer of the Peking–Hankow (often referred to as 'Kin-Han') railway, Thomas Bourne, opened a British government office in Shandong to recruit Chinese labourers to undertake manual work in France.[145]

By April 1917 Bourne had recruited 35,000 workers, and a year later the total number of recruits to the 'Chinese Labour Corps' (CLC) had reached 94,458.[‡] The CLC were very effective at digging trenches, building and maintaining narrow-gauge railways to move supplies, and generally providing logistic support for the soldiers on the front line. They had to be to managed and supervised by Chinese-speaking British

---

* Farrer could not resist this play on words to convey the dual meaning that Will was simultaneously guilty of dereliction of duty towards his country and in bad shape, *i.e.* derelict. A typical example of Farrer's acid wit, the pun is as effective as it is unkind.

† Farrer had received a letter from an almost complete stranger on these lines in Lanchow in September 1915, and had laughed it off. But by mid-1916 he was working for the propaganda unit set up in the Foreign Office by John Buchan, and proudly carrying an ID card which certified that he was not in uniform because he was engaged in work of national importance.

‡ By the end of 1918, the total number of Chinese men who had signed up to the CLC was 140,000.

Nell and her belated birthday present from Will, the red setter 'Mustard' April 1918.

Purdom family papers

officers and non-commissioned officers.* These were in such chronically short supply that by late 1917 the British government was actively searching out missionaries and former missionaries to China and urging them to join up, on the promise that they would be non-combatants. Britain was also recruiting Chinese-speaking college graduates in the United States to work in France on a contract basis.

Bourne had worked with Will on issues relating to the supply of wood and timber for the Kin-Han railway since early 1916. He knew that Will spoke Chinese to a good standard. Will, aged 37 in 1917, would certainly have been conscripted into the armed forces if he had been at home in Britain. Bourne was well over 50, too old to serve, but was 'doing his bit'. It is impossible to believe that he did not suggest to Will that he should volunteer for service with the CLC. Although we don't know in what terms Bourne couched this suggestion, at this stage of the war members of the British public were encouraged to see any male of military age not in uniform as a coward who should be told in no uncertain terms to pull himself together and 'do his duty'. Even if Bourne used more tactful language, it's likely that some other members of the British community didn't pull their punches when making the same point to Will.

---

* Major (later Lt-Colonel) Richard Purdon was one of the CLC commanders: although his name is sometimes misspelled 'Purdom', he was not related to Will.

CHAPTER 16

Will knew that he wasn't a coward, and had proved it when facing down greater dangers than most of his critics had ever encountered, but the combined denunciation and exhortation of the 'white feather brigade' left him depressed and privately uncertain whether he was doing the right thing. Hence, Nell also told Farrer that, "[since] *the end of the war seems so far distant*", if Will decided that there was no opportunity to do his bit for Britain in China it was likely that he would come home: "*at the moment he is only holding the reins, so that no other foreigner may drive, but the stupid old horse does not move at all.*"

The most interesting part of Nell's letter of 20 April 1918, however, is her statement to Farrer that Will was "*at present away from home but just about to return after four weeks on the Peking–Hankow railway, inspecting trees and probable forest land.*" This was simply not true. One of Will's notebooks from this period survives and contains his diary for the four-week trip referred to by Nell.[146] It begins on 12 March and ends when he arrived back in Peking Railway Station at 12.45 on 9 April. On the day Nell wrote to Farrer, Will had been back home in Peking for 11 days. In other words, Nell was deliberately misrepresenting the timing of the trip and Will's whereabouts at the moment she was writing to Farrer.

The only credible explanation for Nell's dissimulation is that she was embarrassed that Will continued to refuse to write to Farrer. She told Farrer, falsely, that Will was away from Peking because that enabled her to pretend that he had not replied to Farrer's latest letter because he had not seen it. At the same time, she told Farrer that Will had frequently begun letters to him and "*often I have asked if they were posted but his only answer was 'I mustn't humour him'. No doubt you will understand and forgive his eccentricity*".

Just what Farrer understood is doubtful, but Will's determination not to engage him in correspondence at this time is quite clear. This may have been partly because Farrer had written to Will more than once to propose that after the war Will should accompany him plant-hunting in Kansu and Tibet in an expedition to be funded by Arthur Bulley. Although Farrer told his friend Augustus Bowles that he had persuaded Bulley that north-west China was the best collecting-area for the expedition and "*Purdom* [should be] *its manager*",[147] this was a proposal to which Will had no intention of acceding.

Will's refusal to write to Farrer also prompts the thought that the relationship between Farrer and Will was, on Will's part, more transactional than Farrer had chosen to understand. This is not to suggest that the friendship between them was not sincere on both sides, but Will definitely had no interest in a repeat performance of his role as Farrer's guide and 'fixer', and may have decided that the least painful way of getting this message across was to keep silent until Farrer worked it out for himself.

The tours of inspection in respect of which Will kept the summary diaries contained in the notebook covered the period 12 March to 9 April (on the Peking–

Hankow line); June 10 to 20 June (travelling around Hunan province); 12 to 15 August (the Lung Hai railway, where Will called at the house of the Belgian railway engineer and self-taught dendrologist Joseph Hers; Hers was out). On 12 October, Will spent a day at Hanyang-chow, near Wuhan, where he had "*a very pleasant time*" over dinner with the Verganni Italian–Romanian family and their friend Mr Forest.* It closes with a brief, unhappy, entry for the period 19 December 1918 to 2 January, 1919, when Will was in Ichang, Hubei Province, "*cold snow right down country.*"†

Apart from this last entry, there is nothing in the notebook to suggest that the writer was experiencing any particular stress or systematic difficulties with his work. The picture which emerges from the brief entries (typically 20 or 30 words) is of a competent forester carrying out tours of inspection along the railway line, with the occasional visit to local places of interest, such as temples or local beauty spots. The sites of nurseries were assessed for suitability and Will frequently noted what species of tree grew well in the area, and their utility for different functions related to building and running a railway. More than once, Will noted the need for urgent planting to stop soil erosion before dangerous washout occurred or recurred. He also interviewed staff or applicants for jobs and noted (sometimes in Chinese) summary details of names and qualifications. On several occasions he mentioned that he was accompanied by his friend and colleague Han An.‡

On 1 June 1918, Frank Meyer disappeared over the side of a Yangtze steamer near Wuhu and was drowned. The American consular authorities noted that prior to his death Meyer had been depressed because of the war§ and ruled out foul play.

Although Meyer may have fallen overboard by accident, most of his friends concluded that he had committed suicide. There is no reference to Meyer's death in Will's diary.

Nor, unsurprisingly, is there anything in the notebook which refers to Will working for the British secret services, although he had written to a school friend in Ambleside saying that he was doing so.[148] In considering just what this may have involved, what cannot be over-emphasised is that Will was most unusual in the expat community in having a sizeable circle of Chinese friends who were agents for change in Chinese society and government. This gave Will an

---

* The Vergannis, father and son, were respectively an engineer working for the Belgian railway company building the Kin-Han line and employed by a Belgian trading company. When I first came across this diary entry, I thought that I had stumbled upon the record of a meeting – it would have been the only one – between Will and the Royal Botanic Garden Edinburgh plant-collector George Forrest. At this time, however, Forrest was confined to his bed by fever in Tsway Tson Tien, on the Lichiang range (Yulong Xue Shan) in Yunnan, over a thousand miles west of Wuhan. I have been unable to identify his namesake.
† I generally have little time for the putative analysis of personality or mood by reference to handwriting, but the cramped, crabbed hand of this entry is redolent of acute unhappiness.
‡ In March 1918, Mrs Han and her baby attended Nell's 29th birthday party in Peking, which suggests a close friendship.
§ The Consul did not know this, but Meyer had also recently heard of the death in Washington of two of his close friends, victims of the 'Spanish flu' pandemic which killed between 20 and 50 million people world-wide, 675,000 of them in the USA.

CHAPTER 16

insight into internal Chinese politics which made him a person of interest to the political officers attached to the British Legation.

Will's access to information about the Chinese rail network would also have been of considerable interest to the Legation during what were very volatile times. The historian Hsi-sheng Ch'i suggests that this period can best be understood by seeing China as a continent not unlike mediaeval Europe, split into rival states ruled by competing warlords very like mediaeval kings, whose authority depended on the loyalty of local armies and constantly shifting alliances with other warlords, and who controlled their domains through family and clan networks, and secret societies.[149] Hence the name by which the period is known in China – 'the warlord era'.

Will with Mustard and Nell, summer 1918.
Purdom family papers

Given that the only way to move armies and equipment with any speed was by rail, a firm grasp of the state of the Chinese rail network, the more detailed the better, was essential to the Chancery (political department) of the British Legation. Will's work gave him an overview of the whole railway system and he could easily find out what was happening at any given time on any individual line. The direct access to the Ministers of Agriculture and Commerce and the Minister of Communications which Will enjoyed* and the insights he gained into their thinking would also have been of interest to the Legation.

Both the Chinese government and the railway companies appreciated the service provided by Will and Han An, and in January 1919 when Will's three-year contract with the Ministry of Agriculture expired, the Ministry of Communication (which was responsible for the railways and worked closely with the foreign companies building railway lines in China) and the Ministry of Agriculture (which was responsible for forests) jointly entered into new contracts with them. The two Ministries agreed to pay their salaries in equal shares and that Will and Han An would divide their time between the Forestry Service of the Ministry of

---

* Will's relationship with Tian Wenlie, Minister of Agriculture and Commerce between December 1917 and February 1920, went well beyond the merely professional. In March 1918, when Will mentioned that his sister was about to celebrate her birthday, Tian immediately said that he would buy her a caged bird as a birthday present, and insisted that Will telephone from his office to ask Nell whether she would prefer a talking bird or one that sang. Nell chose the latter, and Tian personally went to the market to select a bird with fine plumage and a lovely song.

Agriculture and a new Forestry Department to be created within the Ministry of Communication. Neither Ministry wanted to retain Forsythe Sherfesee, who transferred to the Ministry of Finance as a Special Adviser to the Minister.

On 2 February 1919, Will signed a new four-year contract with the two Ministries at a salary of 8,000 Chinese dollars a year. The Chinese currency was backed by silver, rather than the 'gold standard' favoured by most of the rest of the world, and after 1914 the price of silver had soared, making Will's new salary equivalent to over £2,000 p.a.[150]

A major driver behind the new contract was the arch-moderniser Wang Ching Ch'un (in pinyin, Wang Jing Chun)*. Wang had graduated in 1911 from Yale with a degree in civil engineering and in 1912 he was awarded a PhD from the University of Illinois for a thesis on the funding of the construction of the British rail network. He was appointed Deputy Minister of Industry and Commerce in 1912 was responsible for drawing up a set of accounting and administrative regulations which effectively unified the Chinese railway network, which had been built by foreign companies with diverse interests, operating procedures and languages.[151] In 1912 he became Deputy Director of the Peking–Hankow Railway Company and in 1917 Managing Director.[152] Wang's competence and personal integrity won him the respect of both Chinese and foreign colleagues and he and Will became firm friends.

Wang was determined to encourage commercial farming along the line of the railway. He created a special 'seed demonstration train' whose carriages contained an exhibition of modern farming methods and was staffed by expert agronomists who gave lectures to local farmers on improved agricultural methods and on means of bringing crops to market. In parallel to this initiative, the British-American Tobacco Company was promoting the cultivation of tobacco along the line and building facilities to purchase and cure the leaves. Both these campaigns were very successful and they generated profitable freight for the railway company as well as serving Wang's wider ambition to modernise Chinese agriculture.[153]

Against this background, in early 1919 Will and Han An were formally tasked to implement a proposal which they had already presented to the authorities, namely to create nurseries and forestry training schools which would increase China's production of timber so as to reduce the requirement for imports for sleepers, telegraph poles, railway wagons and station buildings. They were also requested to devise and implement a plan which would in the long term make the 800-odd mile Kin-Han line entirely self-supporting in timber.

In carrying out these tasks, Will and Han An benefitted from the fact that that the railway company had already bought three parcels of land for the tree nurseries required by the scheme and were putting up most of the money required. They could

---

* The British and US press often referred to him as C.C .Wang.

CHAPTER 16

Forestry Service officers: Will's friend and colleague Han An is third from the left.
Lakeland Horticultural Society

proceed with confidence, knowing that work would not suddenly have to stop due to lack of funds, something that quite often happened to Chinese government projects.

Although some of the work had been started under the aegis of the Forestry Service, the task facing Will and Han An should not be underestimated. Starting almost from scratch, they had to create an entire system for training foresters and nursery staff, including buying the land on which to build forestry schools, erecting the buildings, writing the syllabus for the training courses, engaging teachers and recruiting students. They had to devise and put in place management structures to run the schools, to monitor the quality of their courses and to manage and to audit their finances. Much the same applied to the tree nurseries and to establishing and managing the teams who would be responsible for planting and caring for the trees when planted.

They were also required to make strategic recommendations about where the need for planting to prevent soil erosion and to protect embankments and bridges was most urgent. They were expected to look forward up to 40 years to decide what kinds of timber would be needed in the future and what species of tree would be best suited to different sites, whereupon the question of where and how to source the millions of seeds required had to be addressed.

All this had to be achieved with, at least initially, very little logistical or clerical support and against a background of confused, sometimes conflicting, responsibilities on the part of local municipal authorities, central government and the (foreign) railway companies. Will's contract included a provision that the railway companies would provide him with a converted railway carriage as a home and a mobile office that could be shifted to where on the line he was needed. He told Nell that since he would henceforth be spending most of his time outside Peking he would be moving out of the rented compound in the West City. Rather than return to Britain, which is probably what Will expected her to do, Nell found a job in the school for missionary children at Kuling, on a mountain-top in Jiangxi Province 1,200 miles south of Peking.

Will and Han An rose to their new challenge with enthusiasm. Within a year, visible progress had been made: several nurseries had been established, planting had begun in three 'forestry stations' along the Kin-Han railway, and

Will's private rail-car and staff, 1919.   Purdom family papers

CHAPTER 16

A 'Forest Station' where seedlings were grown on and forest workers trained.
Lakeland Horticultural Society

forestry schools were turning out workers with the skills required to run both the nurseries and the forest stations. All this had been achieved despite the abrupt resignation of the Minister for Communications, Tsao Ju-lin (in pinyin, Cao Rulin), in May 1919, following the Chinese government's failure to persuade the delegates at the Versailles Peace Conference to return to China the former German 'concession' at Shadong; instead, it was given to Japan. When, on 4 May 1919, this news reached Peking, the students at Peking University organised a protest that turned into a riot in which the home of the pro-Japanese Tsao was sacked and burned: he was lucky to escape over the back wall. The 'May the Fourth Movement' then organised a nation-wide boycott of Japanese goods and mass anti-Japanese (and, to a lesser extent, anti-British and American) demonstrations, but these do not appear to have impacted on the work of what was, after all, a Chinese government agency.

The largest of the Forestry Service stations was 15 miles north of Sinyang (now Xinyang), 900 waterlogged acres in the foothills of Mount Jigong immediately east of the railway line. Will and Han An decided that the American swamp cypress, *Taxodium distichum*, would be suitable for the site, the first introduction of this tree to China. In 1919 they imported and germinated hundreds of thousands of seeds from the USA and in 1920 they and the Forestry Department staff were helped to plant the seedlings on Mount Jigong by the soldiers of the military governor of Honan Province, Feng Yu-Hsiang, 'the Christian General', who was well-known (and well-liked) for requiring his troops to undertake public works when not engaged in martial activities. General Feng is said to have joined in and to have planted with his own hands, near the entrance to the site, what is now a very large London plane, *Platanus x Acerifolia*.

The cypresses thrived and in due course the timber, which is resistant to both rot and insect attack, proved very useful to the railway companies.* The Forestry Service built a handsome Western-style house at Mount Jigong for Han An, who moved there with his family, and he and Will had offices in the colonial-style office-block from which the railway company managed the construction and maintenance of this section of the line.

It was not long before other foreign railway companies started to develop their own forestry programmes, including, for example, the Belgian company building what is now the eastern part of the principal east–west Chinese railway line linking Lanzhou (modern Lanchow) with Lunghai (now Lianyungang), a port on the Yangtze river over a thousand miles away. The company secretary was Joseph Hers, whom Will had first met in 1916 or 1917.

Hers was four years younger than Will but a considerably more important figure in the expatriate community.† A self-taught botanist with a particular interest in trees, he had been horrified by the rate at which Chinese forests were being felled, with no thought of re-planting. Hers planned to plant sufficient trees along the Lunghai line to replace the trees cut down to build it and to meet the future needs of the railway. This required the establishment of four tree nurseries cultivating at any one time up to one and a half million seedlings and cuttings and the planting of more than four million trees along the line of the railway.[154] Hers frequently consulted Will about this work: they became close friends and agreed to collaborate on a book about Chinese trees.‡

Hers also wrote to Professor Sargent in Boston in August 1919 sending him several hundred herbarium specimens, one of which Alfred Rehder identified as a new species of snakebark maple and named *Acer hersii*.§ Although Hers did not mention Will in his letter to Sargent, it may well have been Will who suggested to him that the Arnold Arboretum would be a suitable recipient of his specimens

---

* For many years the Forestry Department also collected swamp cypress seed from Mount Jigong for planting elsewhere in China.
† Hers is now remembered principally through the books of the daughter of a Chinese railway engineer and a Belgian woman, the Chinese author and historian who used the pen-name Han Suyin (her real name was Rosalie Chou). She wrote perceptively about Hers' kind-hearted and generous but at the same time patronising and controlling attitude to China and the Chinese, as well as an amusing account of the terror created in rural Chinese villages by the appearance of the six-foot-six (two-metre) tall Hers with his huge black beard.
‡ It is possible that Will and Hers divided up work on the book by agreeing which trees each of them would write about. An undated notebook of Will's listing 60 Chinese trees, a dozen of them described in detail, the others more or less summarily, survives in the archives of the Royal Horticultural Society. In 1923 Hers published two articles concerning just under 50 trees from northern China. Only *Ginkgo biloba* appears in both lists: Will merely noted the botanical and Chinese names of the tree, whilst Hers wrote about it at length.
§ The modern synonymn of this fine tree is *Acer grosseri*.

rather than the more obvious destination of the National Botanical Garden in Brussels. Sargent returned warm thanks and, ever the opportunist, a request for seeds.

Hers sent some seeds and cuttings to Boston that autumn and in July 1920, probably either on his way back to Belgium for a furlough or on his way back to China, he visited the Arboretum and had a cordial meeting with Sargent, who encouraged him to continue to collect seeds and cuttings and urged him to focus on trees commonly found in Chinese forests.[155] Such species, Sargent complained, had been neglected by most collectors because they devoted disproportionate effort to finding rarities.[156]

At about the same time as his visit to Boston, Hers also visited Kew and over the following seven years he sent to the Arboretum the seeds of some 400 species, including rarities like the beautiful lilac *Syringa pubescens julianae* 'Hers'. Over the same period, he also sent a smaller quantity of seeds to Kew.

Will himself also sent plant material to the Arnold Arboretum. The accession

Will's friend the Belgian businessman and self-taught dendrologist, Joseph Hers, first came to China in 1905 as an interpreter for the Belgian Foreign Service.

Arnold Arboretum Archives

files of the Arboretum record that on 7 and 8 January 1920 (the two dates reflect the days on which the plants were entered into the record and do not preclude their both having arrived at the Arboretum in the same parcel) the Arboretum received one cutting of *Populus cathayana* and one of *Philadelphus pekinensis*. The source of both plants is given as "*William Purdom*" and in the case of the *Philadelphus* there is a note "*collected from the wild*".

Unfortunately, the Arboretum accession records do not always differentiate between plants sent by Will direct from China and subsequent acquisitions of plant material propagated in the United States from plants sent by Will to Boston between 1909 and 1912. *Philadelphus pekinensis* was one of the latter, but neither Veitch or Sargent received a specimen of *Populus cathayana* from Will between

1909 and 1912.* So although the *Philadelphus* cutting received in Boston in 1920 might, at least in theory, have been propagated in the United States from a plant sent by Will a decade earlier, the poplar could only have come directly from China. Its arrival in Boston in early 1920 proves that from late 1919 onwards Will was sending plant material to Charles Sprague Sargent and the Arnold Arboretum.

The two plants received in Boston in January 1920 were followed a year later by cuttings of *Prunus triloba* 'Multiplex' collected by Will from the British Legation garden. He sent a second cutting of the same plant a month later and a rose cutting reached the Arboretum in October 1921.

Will also sent seeds to Kew: a packet containing 12 envelopes of seeds arrived there in April 1920, and a second packet containing three kinds of seeds, including allium seed "*gathered in the Nankow pass*", arrived in February 1921. On 20 October 1921, Kew received from Will a packet of seeds of the Chinese Bush Cherry, *Prunus humilis* cultivated in China for its fruit, the last plant material he collected in China.[157]

These parcels of plants were the first Will is known to have sent to Europe or the USA since April 1916, when he had given Farrer a packet for Sir Harry Veitch containing three envelopes of seeds and a slip of headed paper from the Peking Ministry of Agriculture and Commerce inscribed in a bold, confident, hand "*With the compliments of W. Purdom (more later).*"[158] On 10 June 1916, Farrer delivered the packet to Sir Harry at his home in Chelsea† and Sir Harry made a note of the contents – seeds of pine, birch, and sumac trees – in the back of the ledger of Purdom collections he had retained following the closure of Coombe Wood. It does not appear that Will sent any further plant material to Sir Harry or to anyone else before late 1919. This may in part have reflected the limitations of the wartime postal service, but his 1919 and 1920 despatches of seeds also reflected Will's increasing self-confidence and his desire to raise his personal profile amongst Western botanists and botanical institutions.

One effect of the war had been to blow away much of the petty snobbery which had pertained in British (and other Western) society pre-1914 and which had been a major factor behind the lack of recognition of Will's work and achievements in the 1909–12 and 1914–15 expeditions. Furthermore, after the Armistice the foreign residents of Peking had rapidly put aside the tensions which had arisen between them during the conflict. It is unlikely that after 1918 anyone continued to hold Will's decision not to fight against him, especially given his position as a senior adviser to the Chinese government, a post which put him in the top layer of Peking expat society.

The work of the Forest Service went forward well: the American agronomist John H Reisner estimated that in 1919 about one thousand nurseries were

---

* Will probably did send one, under collection number 672, but that was one of a series of numbers lost in transit.
† Will may have known that Sir Harry had also retained some of the plants and seeds collected by Will between 1909 and 1912 and was growing them in his garden in Slough. The loss of Sir Harry's papers makes it impossible to know whether or not Will sent him further plant material as promised.

CHAPTER 16

established in China (many of them by provincial authorities rather than directly by the Forestry Service), which grew 100 million young trees, and that in the same year between 20 and 30 million trees were planted on one hundred thousand acres of otherwise unproductive land.[159]

Will's personal contribution to what was seen by both the Chinese government and outside observers as a successful ecological initiative which also fostered a much-needed sense of national unity was formally recognised by the Chinese government. On 14 March 1921, the Chinese President formally honoured Will with the award of the medal of the Golden Grain (third grade).[*] On the same day the Minister of Agriculture presented Will with a silver cup and made him a member (fourth grade)[†] of the Order of Rank and Merit.

These were golden days for Will. His path in life had, so far, been neither easy nor comfortable, but he was at last in a good place. He was successfully restoring the lost woodland of China and helping the Chinese to manage their forest resources, both tasks to which he was personally committed and both of which were going forward well. He had, with Hers, made a start on a book about Chinese trees which he could reasonably hope would establish his status in Western horticultural circles as a serious botanist, and he was renewing his links with Kew and the Arnold Arboretum. He would no doubt have liked his salary to be paid more regularly, but it always came through eventually, and was sufficient to maintain a comfortable middle-class lifestyle in Peking (when he was there) and to allow him to indulge his taste for elegant and well-cut clothes.

Will had resumed his correspondence with Reginald Farrer, who in early 1920 had suggested they jointly form a company to import eye-salves into China. (We may suppose that, during the period in 1915 when he and Will were sharing accommodation in Lanchow with the dissolute former patent-medicine salesman Ronald Gilbert, Farrer had taken note of the substantial personal profit Gilbert had realised from selling 'pink pills for pale people'.) Will advised against the idea because in the current troubled state of the country it would be impossible to market the ointment in the rural areas where demand would be highest,[160] and Farrer dropped the scheme. Will was looking forward to seeing Farrer again when he passed through Peking in January 1921 on his way back to England after two seasons spent collecting in Upper Burma.[‡] In October 1920, however, news reached Peking that Farrer had fallen ill and died, alone save for his Gurkha staff, in his camp in the rain-sodden Burmese mountains.

---

[*] Out of nine.
[†] Out of five.
[‡] In 1918, Farrer had pressed Will to accompany him on this expedition, but Will declined, saying that the Forestry Department would not release him from his contract.

Chapter 17
Friendship or love?

On 22 November 1920, a young woman and her aunt by marriage, 'Dolly' McLeod Leveson, disembarked in Shanghai from the steamer *Devanha*, out of Southampton. Dolly came from a family which had been established in Shanghai for decades, had married into another 'Shanghailander' family, and was returning home after an extended holiday in Britain. Her niece, Elizabeth Clifton,* 'Betty' to her friends, was planning to spend four or five months as a tourist in Shanghai and Peking.

Betty was born in January 1900 in Cobham Hall, Kent, a very large and beautiful Elizabethan mansion set in grounds landscaped by Humphrey Repton. Her father, the seventh Earl of Darnley and Baron Clifton of Leighton Bromswold, died before her first birthday, when she succeeded to the Barony (but not the Earldom), becoming the youngest peer in Britain.†

Betty Clifton, aged three, wearing her peer's robes made for the coronation of George V.
Leveson family papers

In 1902, Betty's mother married a senior Royal Navy officer, Rear-Admiral Arthur Leveson. Later the same year, when arrangements began to be made for the coronation of Edward VII, the Lord Chamberlain sent Betty (and her mother) a formal invitation to attend, along with every other peer of the realm.[161] Betty was photographed for the *Illustrated London News,* a blonde cherub wearing appropriately sized peer's robes and coronet, but neither she nor her mother participated in the ceremony.[162]

Betty attended St Leonard's boarding school for girls in St Andrews, Fife, which was founded in 1877 with the aim of providing just as good an education as any boys' school in the country. She sat 'Responsions', the Oxford University

---
* Her full name was Elizabeth Adeline Mary Bligh, Baroness Clifton of Leighton Bromswold.
† Even after she had attained her majority, as a woman Betty could not take her seat in the House of Lords, but from the moment of her father's death she was unquestionably a peeress.

CHAPTER 17

entrance examination, in 1918. Inexplicably for someone who later went on to win first-class honours in the Bar examinations,* she failed. Instead of attending university she 'came out' as a débutante in 1919, by which time the blonde cherub had grown into a tall, dark-haired and dark-eyed beauty with notably elegant long-fingered hands.

In the light of the above, the reader might expect Betty Clifton to have been a thoroughly spoilt young woman with a strong sense of entitlement and little interest in, or sympathy for, persons outside a small aristocratic coterie. In fact, nothing could have been further from the truth. Her contemporaries are unanimous in recording that Betty had

Betty Clifton, 1919.　　　　Leveson family papers

absolutely no 'side', and combined a keen intellect, personal charm and concern for the welfare of others with gaiety and sound judgement. From her teens on, she was never afraid to think for herself, whatever the conventional wisdom on any particular subject. As an adult, her circle of friends included some of the leading intellectuals of the day.

Although during 'the Season' in London Betty had attracted her share of suitors and after January 1921 the male expats of Peking were at her feet, her quiet but firm conviction that a woman should retain her independence after marriage and find personal fulfilment in activities going beyond the role of wife, mother, and chatelaine had so far precluded any serious romantic attachment.[163] This was a state of affairs about which she cared not a whit, but which was perhaps about to change.

Betty spent a month in Shanghai before travelling to Peking to stay with the family of Sidney Barton, the senior political officer at the British Legation, whose wife and two young daughters had travelled out with Betty and Dolly on the *Devanha*. The Bartons prevailed on the British Minister, Beilby Alston, to host

---

* Fewer than one per cent of Bar examination candidates pass with Honours, almost all of them Second Class. Betty must have written a quite outstanding paper on criminal law to achieve First Class honours in that subject.

a fancy-dress ball to celebrate Betty's 21st birthday on 22 January 1921.[164] Betty then settled down to write an "*extravaganza*" entitled *The Moon Flower*, a seasonal pantomime comprising a series of comic sketches and dances in which Sidney Barton played 'Pompostes the Prime Minister of Hoodoo', his wife played the Queen of Hoodoo and their daughter Esmé the Moon Flower. Betty played a Shooting Star, in which role she danced to a Chopin nocturne. Other members of the diplomatic community and their children also appeared, and a good time was had by all at the two sell-out performances in the British Legation theatre on 21 and 22 February, when Betty took repeated curtain calls as author, producer, and stage manager.[165]

The temperature in northern China between December and March is typically around -12°C (10°F). The ground is as hard as iron: any kind of horticultural activity or building work is out of the question. The disadvantages of living in and trying to heat even a well-appointed private rail-car, became very apparent. Will was in the habit of returning to Peking around Christmas-time, where he would spend a few weeks catching up with his paperwork and relaxing with his friends in the British and Chinese communities. Will knew the Bartons socially, and although he did not much like Mrs Barton[166] it is entirely possible that he was invited to the birthday ball or that he was in the audience of *The Moon Flower*. In the small world of the Legation Quarter, he would in any case have met Betty in the course of the Christmas and New Year festive season.

At this time, it was not unusual for British men in their forties, especially those working overseas, whose employers routinely discouraged them from marrying until they had reached the age of 30 or 35, to form romantic attachments with, and to marry, women two or more decades their junior.* Will was just under 20 years older than Betty, but handsome and physically fit. His personal dedication to making the world a better place, starting with the forests and the eco-systems of northern China, must have made a strong impression on Betty, who it is safe to say had never before met anyone quite like him.

It is apparent from Betty's later writings that she procured and read Reginald Farrer's two books about the Farrer/Purdom expedition of 1914/15. In the course of 1921, she also became familiar with Will's personal history to a level of detail which she could not realistically have acquired from anyone but Will himself. For his part, Will could hardly have failed to be attracted by Betty's combination of intelligence, independence of mind and personal charm.

The different strata of British society into which Will, a head gardener's son, and Betty, the daughter of an earl, were born constitute, at first sight, a more serious impediment to any romance between them than does the difference in their ages. To be blunt, none of the 'set' of débutantes who 'came out' with Betty in 1919 would have spared Will a second glance. But Betty's friends would not

---

* In August 1912, for example, Ernest Morrison, then aged 50, married his 23-year-old secretary, Jennie Robin. Betty's parents were married in 1899, when Lord Darnley was 48 and his bride 19.

CHAPTER 17

have been surprised that she saw Will for the man he was and made up her own mind about him. Betty's mother would also have known that there was no point in trying to override her daughter's judgement. Her stepfather, a flag officer in a Service which had since the time of Sam Pepys been implacably meritocratic, might or might not have approved but would certainly have understood. And Will had never in his life accepted being 'put in his place': if he had fallen in love with Betty, he would not have allowed the prospect of the disapproval of other members of society to stand in his way.

It should perhaps be spelled out that the idea that Will might be fortune-hunting could not have crossed Betty's mind; she would have been wryly amused if anyone had suggested it. Cobham Hall and the Darnley estate (which was in any case heavily indebted), were entailed* in the male line and on the death of her father had passed to his brother, Ivo Bligh,† the eighth Earl. Her father's extravagant lifestyle – it is said that *"he spent money like water"*[167] – meant that on his death there was little unentailed capital for his wife and infant daughter to inherit. Betty's mother responded to this situation by removing from Cobham Hall all the furniture and plate, which strictly speaking was not subject to the entail, some of which she then sold. She paid Betty a small annual allowance which Betty supplemented with fees for the articles she wrote for British newspapers and magazines and, in later life, by writing books and plays.‡

In January 1921, Betty decided to prolong her stay in China for a further year, during which she lived with the Bartons in the Legation compound, She made several trips with the family to far-flung parts of the country, including Kashgar on the Mongolian border and in May 1921 to the more traditional tourist destinations of Jehol and the Eastern tombs. When she wrote an account for her own family of the latter trip she described at some length, and in terms which suggest expert briefing, the trees and forests through which the party passed:

"The mountain country was extraordinarily beautiful. Until quite recently it was all thickly wooded in the vicinity of the Tung-ling [the Eastern Tombs], *but Tsao-Kun, the Governor of the province, has been cutting down trees wholesale for the last two years. Coniferous trees of various rare species, mountain oaks and limes were the most general varieties I noted. Wild flowers all the way through the mountains to Jehol were interesting* [...] *Several shrubs I did not know* [...] *After the first two days out* [of Peking], *as we neared Jehol, the country became less wooded ..."*

---

* An entail arises from a family trust which limits the number and class of descendants of the original trustee who can inherit a house or estate, usually in favour of the eldest male heir, who holds a life interest which passes to the next entitled person on his death. It prevents estates being broken up by being divided between siblings.
† A cricketer of renown, he was the captain of the England team which in 1883 brought the Ashes back from Australia.
‡ During her adult life Betty lived quietly but in modest comfort. Her one luxury was overseas travel, mostly to Europe (she would have liked to visit Egypt and the Holy Land but couldn't afford to). On her death in 1937, the probate value of all her assets was just £487.

Betty also joked that the inns in which they stayed *"were infinitely cleaner than we had been led to imagine by lurid accounts from persons well acquainted with Chinese inns before we started on the trip"*. It sounds very much as though before making the trip she had been teased by Will, the only foreigner in Peking who simultaneously possessed extensive experience of staying in Chinese inns and detailed knowledge of the trees and forests of northern China.

One hundred years ago, upper-class young women were carefully chaperoned until they married. Mrs Barton was expected by Betty's mother to keep a close eye on her daughter whilst Betty was in Peking, and there is no reason to suppose that she did not take this responsibility seriously. In any case, in the goldfish-bowl of the Legation Quarter the most trivial indiscretion would have been pounced on as a source of gossip. It is on the face of it impossible that Betty and Will could have moved from being friends to being lovers on anything other than an emotional plane.

And yet ... in September 1921 Betty travelled a thousand miles to the Diamond Mountain district of eastern Korea,[*] where she spent a three-week 'interlude'. She told her family almost nothing about the trip, in contrast to her detailed accounts of visits to Jehol and to Kashgar, but she wrote briefly about it 15 years later in her epistolary memoir *Living in the Country*. Betty went from Peking by train to Mukden (now Shenyang) in Manchuria, thence to Seoul, and implies that she did so alone. She was met in Seoul by someone who had the use of a Ford car to travel to the two Japanese-owned hotels in which she stayed,[168] and who presumably also possessed the language skills needed to manage the three-week stay.

Betty's musings about the trip do not mention her travelling companion or companions, only the *"unusual charm"* of the countryside: *"at that time of year maple trees turn the hill-side into a dendrological aurora borealis, from saffron to dark crimson"*, and that she returned with *"a treasured souvenir"*,[†] a small pine tree[‡] growing in a porcelain bowl.[169] Such an unconventional keepsake would make perfect sense as the memento of a visit to the Korean forests in the company of a forestry expert.

None of Will's personal papers from this period survive, but it appears that some time before November 1921 his Chinese friends learned that he planned to marry.[§] Unfortunately, there is no surviving record of the name of his fiancée.

None of the above is free from ambiguity. *Someone* accompanied Betty on her visit to Diamond Mountain, but there is no direct evidence that Will was the person concerned. Nor is there any direct evidence concerning the identity of the

---

[*] The mountain is now just inside North Korea.
[†] In all her writing Betty was consistently frugal with superlatives: the choice of adjective is significant.
[‡] Probably *Pinus koraiensis*, said by Augustine Henry (in 1906) to be *"very abundant in the Diamond Mountains"*.
[§] Cf. the final paragraph of the memorial stele erected by Will's friends and colleagues (Appendix D). *"It saddens us to think that Purdom lived alone and never married his fiancée."* The group of characters concerned, 卜婚未婚 literally means 'the planned marriage never took place'.

CHAPTER 17

person whom he proposed to marry. Will's social life mostly revolved around his Chinese friends. His fiancée might have been Chinese, which would go a long way to explaining why no one in the foreign community in Peking appears to have known about the engagement.

Alternatively, although the immediate juxtaposition of an expression of sorrow that Will 'lived and died alone' with regret that his planned marriage did not take place strongly implies that it was his untimely death which caused the non-occurrence of the marriage, it is, at least in theory, possible that this part of the eulogy simply records his friends' regret that Will never experienced the joys of married life and left no heirs to enrich his family tree. Those of us who see a possible link between the Korean trip and a planned marriage may be indulging in wishful thinking.

On the other hand, the established facts are consistent with Will and Betty having decided to marry either shortly before or during an unchaperoned joint trip to Diamond Mountain. One hundred years ago, such behaviour inexorably implied that the persons concerned would marry: if it had become known about in Peking, they would both have faced social ostracism if they had not done so in very short order. They may have decided briefly to defer an announcement until after Will had resolved some minor health issues, but Will shared his news with one or more of his Chinese friends, whom he knew would respect his confidence.

It is unlikely that there is further evidence waiting to be discovered which will prove or disprove this hypothesis. There are, however, a few straws in the wind. After Will's death, Betty wrote a poem about him which formed part of an obituary in the *Peking Daily News*:

> To gain dominion over other men
> By arms some men essay; a few desire
> To shape their fellows' actions with a pen;
> Towards commerce' gilded guerdon most aspire:
> This man, more rare, did spend his strength and mind
> Outside an office, trod no barrack-square,
> But made of trees his text-books, and did find
> Much good in flowers, commonest, or rare
> Gentian, which clung to some jagg'd mountain crest.
> For plants he fought Tibetan winds and snow,
> Through hills uncharted wandered in their quest:
> Handling the leaves and petals Asia grew,
> Most expert-fingered patience did he show,
> Learned mid smooth lawns of river-border'd Kew.

Nell Purdom sent a cutting comprising the poem to Will's parents who incorporated it in an 'In Memoriam' notice in the local paper. Will's father pasted the Peking cutting into his scrapbook beside a photograph of Betty. Hers is the only portrait in the album which is not that of a family member.

Shortly after Will's death a memorial stone was installed in the chapel in the British Legation, a simple stone block about the size and shape of a small trunk, or of one of the mule-boxes Will used in his first years in China.* The main inscription on the top of the block is now almost entirely illegible due to erosion, but it concludes with his name and an additional inscription under the central block of text describes him as [a] "*Botanist*". At one end of the block is carved a phrase from Betty's poem, "*Handling the leaves and petals Asia grew*", and the sentence concludes at the other end of the block, "*he made of trees his text book and did find much good in flowers*". The evocation on the monument of the touch of Will's fingers is very effective, and it's hard to imagine that anyone but Betty would have modified part of her poem to achieve it: a third party would surely simply have copied a line or two from the original poem onto the memorial.

Betty also wrote an obituary of Will for the *Journal of the Kew Guild*, setting out in detail his career before he took up the post of Forestry Adviser to the Chinese government and praising his work in that capacity:

> "*our late friend loved his work and kept England's prestige high in a far-off land. No-one can say of Will Purdom that he enriched himself at others' expense, as from all we hear the Chinese Government appears to pay wages very seldom.*"[170]

She concluded with the poem which she had published in Peking.

Readers must make up their own minds about the nature of the relationship between Will Purdom and Betty Clifton. Those tending to intellectual rigour may view the notion of an idyll between them as a sentimental fantasy constructed from piecemeal circumstantial evidence and some fragments of Betty's writings. Persons of a more romantic nature may join me in hoping that Will and Betty shared some happy moments and planned a future together.

---

* The block is just over 18 inches high, one foot wide and 30 inches long.

Chapter 18
A tree falls

In the autumn of 1921, Will suffered a recurrence of the trouble he had experienced when he first came to China, large lumps in his neck. He consulted Dr Jean Bussière at the French hospital in Peking, who recommended an operation to remove them. Will agreed, and they fixed a date in late October.

The surgery went well, but Will contracted an infection for which, in those pre-antibiotic days, there was no effective treatment.[171] He died on 7 November 1921.

Will had not told Nell about what was a minor procedure. The first she heard of his illness was when a telegram arrived at the American School in Kuling to tell her that he was gravely ill. By the time she reached Peking, Will was dead.

The British Consul helped Nell to arrange the funeral, on 9 November. The little chapel in the English cemetery outside the Western gate of the city overflowed with foreign and Chinese mourners. There were so many floral tributes that Nell could not face writing to all those of Will's friends who had sent them, but placed a notice in the *Peking Times* expressing her thanks.[172]

After the service, they laid Will Purdom in a grave under the elms – Peking elms, grafts and cuttings of which Will had sent to Britain at the conclusion of his first expedition to China, less than a decade earlier.

Epilogue

Will's death was front-page news in the Chinese press, both English-language and Chinese. The headline in the *North China Herald* of 12 November 1921 is representative:

"Death of adviser on forestry – Mr William Purdom: Botanist and Explorer: Valuable Servant of the Chinese Government".

The article summarised his career prior to 1916 and concludes

"he planned and administered some excellent forestry work for the Peking–Hankow railway and was recently engaged in a comprehensive forestry survey for the railways of China."

There were brief obituaries in the Ambleside parish magazine and in several local papers but by far the fullest and most thoughtful appeared in the *Lake District Herald* on 26 November 1921.[173] Headlined *"A lover of nature"*, it praised Will's determination to endure dangers and fatigue *"often beyond the borders of civilisation"* in search of *"plants that might perchance be useful to mankind"*, an endeavour which added much to human knowledge.

The obituarist, an old school friend of Will's, recalled many fascinating evenings listening to Will reminiscing about his travels in Tibet before quoting from Farrer's books about the 1914/15 expedition to the effect that:

"the success of the expedition was largely due to Purdom's exact knowledge of the route and of the people whom they met [...] His school-mates will (they cannot help it) feel proud of him and of the success he has attained. Had he lived he would have made a name great in the botanical world, as the man who helped to plant Northern China with trees and shrubs. He was a straight, honest and God-fearing man, a man who would hold out his hand to help any fellow-creature in distress. He would do good wherever he could and never boasted of his kindliness and achievements. Many times in China he carried his life in his hands but he was never heard to complain."

The last three paragraphs bear quoting in full:

"In many ways one could not help but take to Purdom. Once you knew him without reserve, then you knew him for the man he was. Troubles and obstacles were alike nothing to a man of his calibre and determination. He both was strong and gentle, with a wonderful tenderness in his eyes."

Upon one of his expeditions he writes: 'It's great to be away here on the roof of the world watching struggling civilisation and when I look towards the great west at the eternal snows, untrodden by man, and think of the eternal rest in these massive heaps, so mighty one cannot describe them, one wonders if even one human being ever thinks of the great long, long

> *silence after death. I doubt it, for they will go on day after day without much thought of their future. Just so is the majority of humanity'.*
>
> *Purdom had friends all over the world. Wherever he went, and wherever he toiled, he was always a favourite with everybody, a good and true friend. No wonder the Celestials cried when he left them. We shall never know the huge amount of good he did in his rough and kindly way, or how he smoothed over their difficulties and listened to their woes, and why: so that we living may have joy in our hearts over these things that we might possess inviolable for ever."*

A week later, a brief obituary appeared in *the Gardeners' Chronicle*,[174] clearly written by one of Will's Kewite friends, recalling that on his departure from Kew his colleagues had presented him with a watch and chain and a fountain pen, praising the *"rich collection of plants, several of which have a permanent value in British gardens"* he had sent home, and concluding *"Thus, at an early age, is lost to science and horticulture a brilliant young man, one of those pioneers of progress and industry who have established the reputation of their country in distant lands"*.

After the unfair criticism to which he had been subjected by Charles Sprague Sargent, Will would have appreciated the crisply professional tone of the notice of his death which appeared in the *Journal of the Arnold Arboretum*.[175] A short account of his expeditions preceded a summary of his work for the Chinese government, the establishment of the *"now flourishing"* Kin-Han forestry station, and the comprehensive survey of Chinese forests in which he was engaged at the time of his death.

But the bulk of the notice is a list of plants: the four plants named *purdomii* (one of the finest, *Rhododendron purdomii,* had been given that name by Alfred Rehder, the Curator of the Arnold Arboretum herbarium, who was almost certainly the author of the notice); the 20 further plants first introduced by Will and growing at the Arboretum; and finally the flat statement that *"Purdom sent to the Arboretum 550 packages of the seeds of trees and shrubs; and in the Arboretum Herbarium there are specimens collected by Purdom under eleven hundred numbers"*. Rehder's determination to put the record straight could not have been clearer or more effective.

Reginald Farrer's book about the second year of the 1914/15 expedition, *The Rainbow Bridge,* which he had completed in 1918, was published posthumously in October 1921. Like Farrer's book about the first year of the expedition, *On the Eaves of the World,* it was dedicated to *"Bill"*. It did not reach Peking until shortly after Will's death.

After Farrer's death, his mother used scissors to cut out from the pages of his Chinese diaries passages she did not want to be read by others. She also destroyed the manuscript of his last play, and most of his correspondence and personal papers.

EPILOGUE

In late November 1921 Charles Hough received a brief note from Will, posted in Peking a day or two before the fateful surgery, enclosing a packet of seeds of *Prunus triloba*, a four or five-foot-tall variety of almond which in March or April is covered with fragrant pink flowers, followed by bright red fruit. Will praised the beauty of the plant, and suggested where he thought it should be planted in the garden at White Craggs. The letter marked the end of a seven-year correspondence between Will and his friend in the course of which Will had often sent seeds to Clappersgate.

In 1929, when Dr Hough wrote a short guidebook to the garden,[176] by then open to the public, he recounted how, after Will's return from China in 1912, they became close friends, and that

"*I had* [then] *started on the development of the Garden, and he entered with the utmost zeal into the planning and scheming of a landscape worthy of the incomparable setting. All that is of merit in the result obtained is due to his knowledge and foresight.*"

In May 1934, a plaque in the garden was dedicated to the memory of Dr and Mrs Hough and several of their friends and family, including Will Purdom.

On 15 November 1921, the British Legation sent a formal note to the Chinese Ministry of Foreign Affairs expressing regret at Will's death and informing the Ministry that "*all property belonging to this employee of the Ministry of Agriculture and Commerce and the Ministry of Communications falls under the purview of the vice-consul of the British Legation.*"* The Legation asked the Chinese government to transfer all outstanding expenses and salary to the vice-consul as soon as possible "*in order to avoid delay in settling inheritance matters.*"

The Ministry of Foreign Affairs wrote to the other two Ministries concerned to urge them to settle all outstanding payments as soon as possible, but in January 1922 the Vice-Consul sent a further note expressing concern that nothing had been forthcoming. In February 1922, the Consulate received a Ministry of Agriculture cheque, which the bank refused to honour due to insufficient funds in the Ministry account. After a further exchange of notes between the Legation and the Ministry of Foreign Affairs, in August 1922 the Ministry paid the arrears of Will's salary in cash, which the Consulate transferred to Nell.

Betty Clifton returned to England in January 1922 and worked as a journalist and sub-editor at the *Daily Mail* whilst studying for the Bar examinations, which she passed with first-class honours in 1923. In January 1926, she became one of fewer than ten female barristers in Britain. Shortly afterwards, she fell gravely ill. Although in time she made a partial recovery, she lived quietly for the rest of her life, at first with her mother, by then widowed for the second time, in southern

---

* This was an unorthodox request: standard procedure was for the Consulate to ask for the property of the deceased to be delivered to the next-of-kin, so as to avoid the Consul being left 'holding the baby' if there should be a dispute about inheritance. Will's friends in the Consulate may have wanted to be helpful to Nell, or perhaps the Legation wanted to secure Will's papers in case some of them related to his work for the British secret services?

EPILOGUE

Kent and later in Surrey. She wrote magazine and newspaper articles, books and plays, initially under the pseudonym of Lenox Fane. Her circle of friends included Bertrand Russell, Joseph Conrad and Elizabeth von Arnim. Betty died suddenly in 1937. She never married.

Nell Purdom continued to teach at Kuling until August 1922, when she returned to Britain. At the end of the year she took up a post teaching in Singapore. She returned to London in late 1926 and the following year she gifted 120 of Will's photographs from the 1914/15 expedition and some later photographs to the Royal Geographic Society, London. She then took up a position working for the colonial administration in Malaya as Supervisor of Girls' and Women's Education.

In 1936 Nell was appointed the Principal of the first Malay women's training college in Malacca, whose graduates were the first to teach Malay girls in their native language. Later the same year she married George Irving, an Australian surveyor. Unusually for the period, Nell kept her maiden name, and both she and her husband continued to work in Malaya until the Japanese invasion of 1941. Nell was evacuated to Britain from Singapore and her husband was interned there by the Japanese. After the war they returned to Malaya. In 1957, when Malaya gained its independence, they retired to Australia. Ten years later, Nell published privately, in an edition of 35 copies, a memoir of her life with Will in Peking during 1917 and 1918, *Sweet Stone Bridge*. Nell died in 1985, aged 96, the last of Will's six siblings.

Will's father, William, never really recovered from Will's death. He grew in his garden in Ambleside as many specimens of plants introduced by Will to Britain as he could, including the lovely but fiendishly hard to propagate *Rhododendron purdomii*.* When he died in 1933, aged 83, his death certificate gave the principal cause of death as exhaustion. His tombstone records not only his own name and dates of birth and death but also memorialises Will's death in Peking, quoting Isiah 18.11.19, "*I will plant in the wilderness the fir tree and the pine*". Will's mother, Jane, died in 1944 and lies under the same stone.

Annie Groombridge is not known to have had any contact with Will after 1912. She taught for over 30 years in an elementary school in Buckhurst Hill, Essex, retired in 1949 and died in 1967.

Ernest Morrison continued as special adviser to the Chinese Government including after the death of his original patron, Yuan Shih K'ai. He worked to bring China into the war on the Allied side, which happened in August 1917. Morrison visited Australia in late 1917 and campaigned unsuccessfully for the introduction of conscription and for closer ties between Australia and China so that a strengthened China could challenge Japanese expansion.

---

* Writing in the *2021 Camellia, Magnolia and Rhododendron Yearbook*, Esders, Ridderlöf and Salomonsson have challenged the identification as *Rh. purdomii* of the plant grown by William, suggesting that it may have been another rhododendron species collected by Will on the Taibai Shan.

EPILOGUE

In 1918 Morrison returned to Peking and was appointed a member of the Chinese delegation to the Paris Peace Conference. Despite being seriously ill, he attended the early sessions of the conference but in May 1919 he was forced to go to Britain for medical treatment. He died there in May 1920. Sad to say, one of his last letters, to his old friend Sir John Jordan, the former British Minister in Peking, complained that his salary from the Chinese government was inadequate, especially when compared to that of *"a gardener like Purdom"*, whom he considered to be grossly overpaid.

In 1923, the *Société Dendrologique de France* published two papers by Joseph Hers on northern Chinese trees, the first on willows and poplars and the second on conifers. Both papers read very much like chapters from, or draft chapters for, a larger work, and the style and length of Hers' descriptions of individual species is similar to those contained in the draft outline written by Will.* Hers refers several times to information received from Will and to 'having very often compared notes with my late lamented friend, Purdom'.[177]

Taken together, Hers' 1923 publications and Will's draft outline offer a tantalising glimpse of what a Hers/Purdom book on Chinese trees might have looked like: a clear and readable manual for professional foresters and arboriculturists giving an account of the distribution, history and traditional uses of each tree, with detailed advice on its cultivation and utility, especially for anything connected with railways, the whole illustrated with good-quality photos (like Will, Hers was a skilled photographer).

In 1924, Hers discovered that the Board of the Lunghai-Pielo Railway Company intended to apply the whole of the year's profits to the extension of the line, in breach of their loan agreements with Western banks who were entitled to have interest due to them paid first. He made the Board honour the terms of the loans and the Chinese government responded by forcing his resignation. Hers was then engaged by the influential *Société Belge d'entreprise en Chine* (of which, in time, he became Secretary and, later, Director). The company represented the Belgian banks financing various enterprises in China. Hers moved to the firm's Hankow office, and later to the Shanghai HQ. His long trips through the botanically unexplored country along the line of the railway ceased and after 1925 he did little plant-collecting.

Hers enjoyed considerable success as a businessman in China until 1938, when he went to Belgium for a period of leave and found himself unable to return to China because of the outbreak of the Second World War. He lived in Belgium until his death in 1965, aged 81, when most of his collection of over 6,000 herbarium specimens was divided between the Belgian National Botanical Gardens and the Arnold Arboretum.

---

* Hers' articles were of course written in French, although he subsequently published English translations in other journals.

EPILOGUE

Sir Harry Veitch continued to play an active role in the Royal Horticultural Society after winding up James Veitch & Sons. He also sent seeds from his garden at East Burnham, Slough, for trial at the RHS gardens at Wisley. After the death of his wife in 1921, he retired to Slough and died there in 1924, aged 87.

Charles Sprague Sargent died in March 1927, one month short of his 86th birthday. He had been the Director of the Arnold Arboretum for over half a century but had made no attempt to groom a successor. A somewhat confused and unhappy period followed before Professor Oakes Ames of Harvard College took over, initially as Supervisor and later as Director of the Arboretum.

Ernest Wilson continued to be lionised by the American press and public and wrote seven books and numerous articles about his expeditions. Following Sargent's death, he was briefly in charge of the Arnold Arboretum. Wilson was disappointed not to be appointed Director, but Oakes Ames made him Keeper of the Arboretum. In 1930 Ernest Wilson and his wife died when he lost control of his car, which rolled down an embankment.

Forsythe Sherfesee continued as Special Adviser to the Chinese Ministry of Finance until 1927, when at a reception in Peking he met Emily Ryerson, a wealthy American who had lost her husband in the *Titanic* disaster. She was 19 years his senior and touring the Far East. They married later that year and travelled widely before settling on the French Riviera, where they lived in some style and made an important collection of 1930s furniture which is now in the Smithsonian Museum. Emily died in 1939 and Forsythe in 1971.

Prince Yang Chi-Ching, the despotic hereditary ruler of Choni, hosted the Austrian–American botanist and plant-hunter Joseph Rock during 1925 and 1926 and provided him with a military escort so that he could botanise in the region around Choni. Rock was distressed by Yang's cruelty and rapacity. Writing years after he had left the region, he claimed that in 1928 Yang raped and killed four women, the daughters-in-law of General Ma Fu-hsiang (pinyin Ma Fuxiang), whom his soldiers had captured in an attack on a government column. Ma retaliated by burning and sacking Choni and the adjacent monastery and slaughtering the inhabitants. Yang managed to flee, but according to Rock was killed by his own officers and servants when he returned and tried to re-establish his rule.

Rock's narrative is difficult to reconcile with the official Chinese government account of Yang's death, which is that in September 1935 he helped the Communist forces as they withdrew west on the Long March by providing information and guides to enable them to outflank and defeat Kuomintang forces in the Lazikou Pass, the last natural obstacle before the marchers' final destination in the far west of Kansu. Yang and his family, save one son, were executed by the Kuomintang in reprisal for this action. A large monument outside Yang's former palace in Zhuoni reprises the above and declares Yang to be a martyr of the Communist Revolution.

In 1922 Han An was appointed Director of Educational Affairs and Head of the Chinese Department of Forestry. In 1934, after a period of service in Hebei and Anhui provinces, he was appointed Head of Forestry for Shansi Province and in 1941 he became Head of the Central Forest Laboratory. He retired to Xi'an in 1949. Han died in 1961, aged 78, and is remembered as one of the leading pioneers of Chinese scientific forestry.

In October 1922, Will's friend from the Ministry of Communications, Wang Ching Ch'un, by now Director General and Chairman of the Board of the Chinese Eastern Railway, led a 45-person delegation comprising staff from the Ministries of Agriculture and Communication, members of the Forestry Service and botanists from Peking University to Mount Jigong Forestry Station, near Sinyang (now Xinyang). There they formally re-named the Forestry Station the Purdom Forest Park and dedicated a stele to honour the memory of Will Purdom.

The inscription on the stele records Will's career in Britain and China.* It praises his personal integrity and love of China, his knowledge of botany and his diligent and meticulous work of reforestation.

"*He was a loyal and true friend of the Chinese* [people] *and won the admiration and respect of his colleagues* [...] *If only he could have lived to a hundred years of age, he would certainly have trained the next generation of forestry experts.*"

The last paragraph of the inscription, a eulogy probably written by Wang Ching Ch'un, reads

"*Purdom worked tirelessly for the reforestation of China. He was a virtuous and upright man. We are saddened that our friend lived and died alone and never married his fiancée. His family members were thousands of miles away at the time of his death. There is one thing we can be sure of: Purdom's spirit will roam either the mountains and valleys of north-western China, or the vast forests along the Kin-han railway.*"

Almost all the trees planted by Will on Mount Jigong were felled in 1942 during the Japanese occupation of Henan Province, but the site was replanted after the 1948 Revolution and the 89 survivors of Will's planting – all of them by now fine specimens – are carefully protected.

The Purdom Forest Park is a popular local amenity and is maintained to a high standard by the Xinyang Municipality. In May 1990, concerned that Will's memorial stele was showing signs of erosion, the park authorities erected a four-pillared pavilion over it. At around the same time, they created a visitor centre in the former offices of the Forestry Station, which features a series of paintings depicting Will's life and that of Han An both before and during Will's time in China. The paintings are naïve in style and the captions unsophisticated, but it is apparent that in Xinyang Will is remembered with affection.

---

* The full translation is at Appendix D.

EPILOGUE

Wang Ching Ch'un, who as deputy Minister of Industry and Commerce worked with Han An and Will to restore forests and develop commercial agriculture along the Kin-Han line, played a key role in in establishing an effective national Chinese railway service. Between 1931 and 1949 he was Head of the Chinese Government Purchasing Commission in London, responsible for buying materials for construction and maintenance of rail lines in China, steamships, and telecommunications equipment. The London Purchasing Commission closed after the Communist Revolution, and Wang moved to California. On his death in 1956, the modest size of his estate was cited as conclusive confirmation of his lifetime reputation for integrity in the discharge of his public duties.[178]

All the memorials in the British Legation chapel were removed from the building when the Chinese government nationalised it in 1957. Those (the majority) not claimed by the families of the persons commemorated are now in the garden of the British Ambassador in Peking. In 1999, the plantswoman and garden designer Jan Galsworthy, the wife of the Ambassador, procured from the Royal Botanic Garden Edinburgh a specimen of *Rodgersia aesculifolia* 'Purdomii' (previously known as *Rodgersia purdomia*) and planted it next to Will's memorial, where it continues to thrive.

The English cemetery in Peking was nationalised by the Chinese government in 1951, but although the authorities developed the adjacent fields, they initially left the cemetery alone. In 1966, Mao Tse Tsung, determined to destroy old ways and old thinking, unleashed on China the Cultural Revolution. In August 1967, the paramilitary 'Red Guards', students and young people licenced by Mao to use violence to bring about change, devastated the cemetery, smashing tombstones, cutting down trees and destroying the chapel. They also burned down the British Embassy. The site of the cemetery is now beneath the Second Peking Ring Road.

The *Aesculus chinensis* seedling Will brought back to London in May 1912 and which found its way to Kew is now a handsome tree, 50 feet tall and with a trunk 20 inches in girth.[179]

List of plates

Front cover: Will in the grounds of the Temple of Heaven, Peking, in April 1912.

List of mono plates	Page
1. Brathay Hall from Lake Windermere	1
2. Purdom family	2
3. Brathay Lodge	2
4. Harry Veitch	14
5. Will, aged 21	15
6. William Thiselton-Dyer	22
7. David Prain	33
8. Charles Sprague Sargent	40
9. Will self-portrait, Peking 1909	66
10. On the Luan river	69
11. The Great Wall from the Luan river	69
12. Camp in the Wutai-Shan	70
13. Will's carts and mules	71
14. Will with priests, Wutai-Shan	73
15. Tibetan peasant girl	76
16. Will with inn-keeper and family	80
17. Will and the Mafu at a gate in the Great Wall	80
18. Will's camp, Moutan-Shan	84
19. Will on horseback	85
20. Deforestation in Shansi Province	86
21. Summit of the Qinling mountain range	87
22. Prince Yang Chi-Ching, ruler of Choni	95
23. Choni town	97
24. Will with armed escort	98
25. Acho-che-ro, reformed robber chief	99
26. Fenwick-Owen and Wallace departing Lanchow	103
27. Escort of Muslim soldiers	104
28. *Aesculus chinensis* in a temple in the Western Hills	106
29. Group in the Temple of Heaven, Peking	107
30. Frank Meyer	121
31. White Craggs, c. 1910	124

List of mono plates	Page
32. Carts bogged down on the road to X'ian	135
33. Will in camp	137
34. Will's staff, Ma and Lay-go	140
35–38. Will disguised as a muleteer	141
39. Farrer's room in Lanchow	147
40. Will, Reg Farrer and Zhang Bing Hua, Lanchow	148
41. Mule-train above Wolvesden House	157
42. Wolvesden House	157
43. The yard at Wolvesden	158
44. Looking down on Wolvesden House	158
45. Tea party, Peking	169
46. Will's study in Peking, 1917	170
47. Nell and Mustard	173
48. Will, Mustard, and Nell	176
49. Forestry Service Officers, including Han An	178
50. Will's private rail-car	179
51. A 'Forest Station'	180
52. Joseph Hers	182
53. Betty Clifton in coronation robes	185
54. Betty Clifton, aged 19	186

Notes:

Plates 13, 25, 28, 30 and 52, from the archive of the Arnold Arboretum, are reproduced by kind permission of the President and Fellows of Harvard College.

Plates 9–27, 29–45 and 47–51 are photographs taken by Will Purdom, some of them using a clockwork self-timer.

Plate 28 is a photograph taken by Joseph Hers.

Maps

Will Purdom's area of activity 1909–12.	72
Will Purdom and Reginald Farrer's area of activity 1914–15.	143

LIST OF PLATES

List of colour plates   **Page**

I. A Sanderson "Tropical" camera   205
II. Harebell poppy   205
III. The 1910 herbarium specimen of *Rhododendron purdomii*   206
IV. *Rhododendron purdomii* on the Taibai-Shan   207
V. Farrer watercolour, 'Lonely Poppy'   208
VI. Purdom Forest Park, Xinyang   209
VII. Will's memorial stele, Xinyang   210
VIII. Will's British Legation chapel memorial   211
IX. *Aesculus chinensis*, Kew   212

LIST OF PLATES

A Sanderson "Tropical" camera, as used by Will in China.　　　　　　　　　　Alamy

Harebell poppy at Ingleborough Hall, 1950s, perhaps a descendant of the stock sent back by Farrer in 1915.　　　　　　　　　　Royal Botanic Garden Edinburgh

LIST OF PLATES

The 1910 herbarium specimen of *Rhododendron purdomii* held by the Arnold Arboretum. This is the 'type' specimen, which is literally definitive: if another specimen matches this one, it is *Rh. purdomii*, otherwise it isn't.

Arnold Arboretum Archives

Rhododendron purdomii at 11,000 feet (3,400m) on the Taibai-shan.

Sten Ridderlöf

LIST OF PLATES

Reginald Farrer watercolour of the 'Lonely Poppy', *Meconopsis psilonoma*, August 1914.
Royal Botanic Garden Edinburgh

Purdom Forest Park, Xinyang: it is obvious why Han An and Will decided that the American swamp cypress, *Taxodium distichum*, was well suited to this site.
Cheng Yi Meng

LIST OF PLATES

Will's memorial stele, Purdom Forest Park, Xinyang.   Cheng Ying

The memorial to Will installed in the British Legation chapel shortly after his death, now in the garden of the British Ambassador. *Christina Scott*

LIST OF PLATES

The *Aesculus chinensis* Will brought to London as a small seedling in 1912 and which Sir Harry Veitch passed to Kew.
Francois Gordon

Appendix A

Paper on afforestation written by Will Purdom in Spring 1912 and handed by him to Ernest Morrison

THE AFFORESTATION QUESTION IN CHINA

Extensive travelling in the Northern Provinces of China on Forestry and Botanical research has borne in upon me deeply the necessity for a scheme of afforestation. The present time seems ripe for pointing out a few of the drawbacks attending the wanton destruction of what must have been fine old forests.

The influence of the loss of these forests on the climatic conditions of the country need hardly be pointed out, though this must be serious. The effect, however, on the general contour of the land from erosion, the depredations made by extension of river-beds from silting and the exposure of crops to full climatic conditions should engage the serious attention of those interested in the welfare of the agricultural population.

Fuel is notably scarce in China, but even this does not excuse the wholesale scraping from mountain sides of every particle of vegetation, a matter of compulsory daily occurrence in many provinces which results in the extinction of all soil-binding plants, to say nothing of the many fine species of trees and shrubs thus lost to the world. This havoc is particularly galling when one thinks of the effect the development of the country's immense coalfields would have on this question of fuel.

Here and there remnants are to be found of past forests, consisting mainly of pine, fir and larch and judging from some of the specimens seen, excellent timber for almost any purpose must once have been available. Particularly in the Weichang district in Chili is this apparent. There appears to be in that neighbourhood neither [a] system of felling trees nor any restriction as to amount. Thousands of fine poles were observed rotting in the grass, poles too of a size suitable for holding telegraph-wires. In many parts of Shansi, Shensi and Kansu similar denudation exists but there is no reason why, with organised effort, many parts, now barren waste, should not be again afforested.

Between Jehol and Weichang, too, erosion is hard at work; deep gullies forming in the hillsides every year, the soil and sand thus displaced finding their way to the valleys, widening out the river-beds, hiding all trace of roads in places, and making cultivation of crops an impossibility, except on the slopes of the hills.

Apart from all these considerations, the lack of timber will seriously affect the development of the country in such matters as extension of railways, development of mines and building of bridges, both from the point of view of cost and facility of carrying out these enterprises. Many small local industries, e.g. the table-making in Jehol and the ornamental work in Walnut at Tainchow in Kansu must in time suffer from want of material, though, beside the main issue, this aspect of the matter is comparatively unimportant. There is, however, another serious factor in the denudation of land which will one day have to be considered – the encroachment of sands from the northern deserts. Whilst on a visit to Yulinfu in North Shensi, the writer was particularly struck by this. For many *li* south of the Great Wall sand prevails, choking out all signs of vegetation. Here the planting of soil-binding plants would prove of the greatest benefit, and even further south on the great Peking plain the same policy could be used with a view to protecting the region from the heavy winds which sweep across it in spring.

On this deeply interesting subject of afforestation, so far-reaching in its effect, so bound up with commercial and industrial prosperity, much more might be said. These few words are written, however, in the hope that those who in the future direct the agricultural policy of China will seriously consider the advisability of tackling the question, particularly as China contains within itself adequate and excellent stock from which to make a beginning.

Faithfully yours
William Purdom

Appendix B

Suggestions for a forestry training school sent by
Will Purdom to Ernest Morrison in July 1913, with an album
of photographs showing the effects of deforestation.

SUGGESTIONS FOR TRAINING FOREST OFFICERS

As will readily be seen the photographs show three things, viz:-

That at one time, quite a large proportion of country in the north was under virgin forest.

That great devastation has taken place, with denudation of soil, and the choking up of rivers, these having ultimately occupied and flooded whole valleys.

That there still remain remnants which, if protected, would provide stock for future operations.

As evidence of the third fact, I would point out, that from seeds harvested from remnant Pines at Jehol and Weichang, we have now growing in England many thousand seedlings. The same applies to the fine Birch, Larch, and Spruce of Western Kansu.

Now, although it would appear a difficult problem to start, China being confronted as it is with other important matters, there are many preliminary points which could, even now, be proceeded with.

A forest school might be established in the vicinity of Peking, say in one of the large temple courts. Young Chinese students could be prepared for acting supervisors in the future. They could be given a grounding in Theoretical and Practical Forestry.

Direct and indirect utility of Forests.

Formation and Regeneration of Woods.

Harvesting of seeds, sowing, planting, protection of woods, forest laws, and general management.

In addition, a second batch of working foresters could be trained in practical forestry operations, viz:- sowing, planting, and tending of restocked areas. These men would be the leading workers under the direction of the supervisors. If trained, they would be in a position to direct coolie labour, when planting was taken up to any large extent.

These few ideas are set forth as a step to the larger issue. The very serious facts of treeless wastes with the accompanying evils of erosion and floods, the scarcity of wood as a raw material for the people, all tend to stir up discontent and therefore bring great anxiety to the government. I believe those responsible for the administration of the country would be well advised to immediately take steps to facilitate the establishment of such a training school, and get under way a group of men versed in the principles of forestry, its advantages and possibilities, thereby assisting the Empire into prosperity.

W. Purdom

Appendix C

Prose poem about plant-hunting on the Tibetan border in 1915 written by Will Purdom in December 1916

What about a few Kinlingkiang Cakes?
The dear old Hut at Wolvesden?
The Chago Valley in May or September?
The great weird craggs of Minshan?
The Tan-kwei and his pig?
The sheep at Tientang?
or
The poppies in Mirgo or Datung?
No more by God.

Appendix D
Transcript and translation of stele at Purdom Forest Park, Xinyang

维中华民国十年十一月七日,农商部、交通部公聘客卿林务员?人波尔登卒于京师。越二日,葬于西便门?国坟地。凡我同人,永怀哀悼,靡所置念,相与推先生德?,记述于碑??????

先生研深植学,来游中华者十数年。考树性、辨土宜,凡西北高山广漠大荒之间,足迹殆遍,心有所得,?笔于书。民国四年,农商部聘襄林政。又三年,交通部亦聘之。

先生勤于事,详于物,忠诚而和蔼,友于异国,情同昆季,僚(穴采)亦爱慕焉。

若论客之德行,实若余?邵豹?于?也。倘假寿百年,不仅足于树木,且足树人矣。

惜功业未半,中道崩殂,得年三十有七。?夫!铭曰

??波尔登??于?用造乎中国之林,劳足以摇精,德足以感人。悼君孤?,卜婚未婚,悲君骨肉海阻,云?胡天不吊名莫能名。知君之魂纵不来往于西北数万里之高山大壑,亦必萦扰于京汉路旁数千兆之松柏与蓁苓。

五等嘉?章湖北任用县知事胡宗焕撰文

四等嘉禾章国务院存记道尹?国勋书

??任用县知事安徽太湖?布?刻石

中华民国十一年岁次壬戌年十月立

Translation

On 7 November 1921, our esteemed foreign colleague William Purdom, who was employed by the Ministry of Agriculture and Commerce as well as by the Ministry of Communications, passed away in Peking. Two days later, he was buried in the British cemetery at Xibianmen. His grieving friends resolved to erect this stele as a permanent memorial to his character and virtues.

Purdom carried out extensive botanical research and travelled around China for over a decade. He studied the characteristics of various tree species and examined the different soils of China. He visited the towering mountains and vast deserts of northwest China and made written records of his findings.

In 1915, The Ministry of Agriculture and Commerce employed him to manage forests. Three years later, the Ministry of Communications also employed him.

Purdom was most diligent and meticulous in his research. He was a loyal and true friend of the Chinese people and won the admiration and respect of his colleagues.

Speaking of his virtues, [illegible due to weathering]. If only he could have lived to 100 years of age, his achievements would not have been limited to planting trees: he would certainly have trained the next generation of forestry experts. His tragic death at the young age of 37 [sic], leaving so much unaccomplished, fills us with sorrow.

APPENDICES

William Purdom worked tirelessly for the reforestation of China. He was a virtuous and upright man. It saddens us to think that Purdom lived and died alone and never married his fiancée. His family members were thousands of miles away at the time of his death. In any event, we are certain that Purdom's spirit will roam either the mountains and valleys of north-western China or the vast forests along the Kin-Han railway.

Composed by: [illegible] Fifth grade Hu Zonghuan
Calligraphy by: [illegible] Fourth grade [surname illegible] Guoxun
Carved by: [title and name illegible] of Taihu, Anhui
[there follow 45 Chinese names, the first of which is Wang Ching Ch'un]
Erected in October 1922

Sources

In January 2018, the RHS magazine, *The Garden,* kindly published a letter from me concerning Will Purdom, and appealing for members of the Purdom family to contact me.

To my great good fortune, Will's great-nephew, Alan Purdom, did get in touch. I also received an email from Mr Alistair Watt, the Australian botanist, plant-hunter and biographer of Robert Fortune, who told me that he was working on a book about Will Purdom and Reginald Farrer. He and his wife had recently followed the route taken by the two men through western China in 1914/15 and he hoped to publish in 2019.

No-one is really happy to learn that someone else is also working on a book about the person he is researching, but it was apparent that Mr Watt and I were approaching Will from different angles. He was largely focussed on the plants Will and Farrer collected and sent back, whereas I was primarily concerned with the political and social background to Will's life and with the personal choices he faced in the course of it.

Mr Watt published *Purdom and Farrer, Plant Hunters on the Eaves of China,* in early 2019. Not wanting my view of Will's personality to be influenced by Mr Watt's, I held back from reading his book until the end of the year, after I had completed my own manuscript.

A further biography of Will, *A Perfect Friend, the life of Cumbrian plant hunter William Purdom,* by Victoria Presant, was published by the Hayloft Press in late 2019. This looks at Will's Westmorland antecedents and background and offers some interesting insights into how these affected his personal experiences in China and elsewhere.

Bibliography

Bickers, Robert *The Scramble for China: Foreign Devils in the Qing Empire 1832–1914*, Allen Lane, 2011.

Bickers, Robert *Britain in China*, Manchester University Press, 1999.

Bickers, Robert *Getting stuck in for Shanghai: the British at Shanghai and the Great War*, Penguin, 2014.

Bickers, Robert *Empire Made Me,* Allen Lane, 2003.

Bishop, George *Travels in Imperial China: the exploration and discoveries of Pere David*, Cassell, 1990.

Boorman, Howard (editor) *Biographical Dictionary of Republican China*, Columbia University Press, 1970.

Bowles, E A *My Garden in Spring*, T C & E Jack, 1914.

Boyd, Julia *A dance with the dragon: the vanished world of Peking's foreign colony*, I B Tauris, 2012.

Bredon, Juliet *Peking, a Historical and Intimate Description of its Chief Places of Interest*, Kelly & Walsh, 1922.

Bretschneider, Emil *History of European botanical discoveries in China*, DDR Zentral-Antiquariat Leipzig, 1981 reprint of original 1898 edition.

'British Resident' *British Memorials in Peking*, Tientsin Press, 1927.

Bulag, Uradyne E *History and the Politics of National Unity*, Rowman & Littlefield, 2002.

Bridge, Ann *Peking Picnic*, Chatto & Windus, 1932.

Briggs, Roy W *'Chinese' Wilson, a life of Ernest H Wilson*, HMSO, 1993.

Brockway, Lucille H *Science and Colonial Expansion, the role of the British Royal Botanic Gardens*, Yale, 2002.

Chang, Jung *Wild Swans: three daughters of China*, Harper Collins, 1991.

Chang, Jung *Empress Dowager Cixi*, Jonathan Cape, 2013.

Chapman, Phil *Wild China*, BBC Books, 2008.

Charlesworth, Michael *The Modern Culture of Reginald Farrer*, Legenda, 2018.

Ch'en, Jerome *Yuan Shih-k'ai*, George Allen & Unwin, 1961.

Chesneaux, Jean *China from the Opium Wars to the 1911 Revolution*, Pantheon Books, 1976.

Chesneaux, Jean *China from the 1911 Revolution to Liberation*, The Harvester Press, 1977.

Chu, Samuel C *Reformer in modern China: Chang Chien 1853–1926,* Columbia University Press, 1965.

Clifton, Baroness *Poems*, Spottiswoode Ballantyne, 1934.

Clifton, Baroness *Living in the Country*, Country Life, 1935. (For Baroness Clifton, see also Fane, Lenox.)

Cohen, Paul A *Discovering History in China*, Columbia University Press, 1984.

Corbett, B O *Annals of the Corinthian Football Club*, Longman, 1906.

Cowan Macqueen, J *The Journeys and Plant Introductions of George Forrest VMH*, RHS/OUP, 1952.

Cox E H M *Plant-hunting in China*, OUP, 1945.

Cox, Kenneth *A Plantsman's guide to Rhododendrons*, Ward Lock, 1989.

Crook, J Mordaunt *The Rise of the Nouveaux Riches: Style and Status in Victorian and Edwardian Architecture*, John Murray, 1999.

Cunningham, Isabel Shipley *Frank N. Meyer: plant hunter in Asia,* Iowa State University Press, 1984.

De Charms, Leslie *Elizabeth of the German Garden,* Heinemann, 1958.

Desmond, Ray *History of Kew,* Harvill/Kew, 1995.

Desmond, Ray & Hepper, Nigel *A century of Kew Plantsmen,* Kew, 1993.

Drayton, Richard *Nature's government: Science, Imperial Britain, and the 'improvement' of the World,* Yale, 2000.

Endersby, Jim *Orchid: a Cultural History,* Kew, 2016.

Fan, Fa-ti *British Naturalists in Qing Chin,* Harvard, 2004.

Fane, Lenox (pseudonym of Betty Clifton), *Legation Street,* Little Brown, 1925.

Farrer, Reginald *Among the Hills,* Edward Arnold, 1911.

Farrer, Reginald *On the Eaves of the World,* Edward Arnold, 1917.

Farrer, Reginald *The English Rock Garden,* TC & E C Jack, 1919.

Farrer, Reginald *The Rainbow Bridge,* Edward Arnold, 1921.

Fearnley-Whittingstall, Jane *Peonies, the Imperial Flower,* Weidenfeld, 1999.

Felton, Mark *The British Military in the Middle Kingdom 1839–1997,* Pen and Sword, 2013.

Fenby, Jonathan *The Penguin History of Modern China: the Fall and Rise of a Great Power,* Allen Lane, 2008.

Fenby, Jonathan *The Dragon Throne, dynasties of Imperial China 1600 BC–AD 1912,* Quercus, 2008.

Fergusson, W N *Adventure, Sport and Travel on the Tibetan Steppes,* Charles Scribner's Sons, 1911.

Fisher, F H *Reginald Farrer: Author, Traveller, Botanist and Flower Painter,* Alpine Garden Society, 1932)

Fisher, John *British Diplomacy and the Descent into Chaos: The career of Jack Garnett,* Palgrave Macmillan, 2012.

Fitzherbert, Margaret *The Man who was Greenmantle,* Murray, 1983.

Flanagan, Mark & Kirkham, Tony *Wilson's China A Century On,* Kew, 2009.

French, Paul *Midnight in Peking,* Penguin, 2011.

Glover, Harrell, McKhann and Swain (editors) *Explorers & Scientists in China's Borderlands 1880–1950,* University of Washington, 2011.

Goodman, Jim *Joseph F. Rock and His Shangri-La,* Caravan Press, 2006.

Grantham, G and MacKinnon, M (editors) *Labour Market Evolution,* Routledge, 1994.

Han Suyin *The Crippled Tree,* Cape, 1965.

Han Suyin *A Mortal Flower,* Cape, 1966.

Hansen, Eric *Orchid Fever,* Methuen, 2000.

Hattersley, Roy *The Edwardians,* Little Brown, 2004.

Hay, Ida *Science in the Pleasure Ground: A History of the Arnold Arboretum,* Northeastern University Press, 1995.

Heffer, Simon *The Age of Decadence: Britain 1880 to 1914,* Random House, 2017.

Heriz-Smith, Shirley *The House of Veitch,* St Bridget Nurseries, 2002.

Hickman, Bronwen *Mary Gaunt, Independent Colonial Woman,* Melbourne Books, 2014.

Hinkley, Daniel J *The Explorer's Garden,* Timber Press, 1999.

Hoare, J E *Embassies in the East,* Curzon, 1999.

Hosie, Lady *The Pool of Ch'ien Lung,* Hodder, 1944.

Hosie, Lady *Two Gentlemen of China,* Seeley Service, 1924.

BIBLIOGRAPHY

Hough, Charles *Henry A Westmorland Rock Garden*, Frederic Middleton, 1928.

Hui-Min, Lo (editor) *The correspondence of G E Morrison 1895–1912*, vol. 1), Cambridge, 1976.

Hui-Min, Lo (editor) *The correspondence of G E Morrison 1912–1920*, vol. 2), Cambridge, 1978.

Hunt, Peter (editor) *The Golden Age of Plant Hunters*, Dent, 1968.

Illingworth, John & Routh, Jane *Reginald Farrer Dalesman, Planthunter, Gardener*, University of Lancaster, 1991.

Jackson, Joe *The Thief at the End of the World: Rubber, Power, and the Seeds of Empire*, Duckworth Overlook, 2008)

Johnson, Margaret *The life and times of Dr J A C Smith 1873–1929*, privately printed, 2011.

Johnston, Reginald *Twilight in the Forbidden City*, Gollancz, 1934.

Jones, Ian *The House of Hird*, Jones, 2002.

Jowett, Philip *China's Wars*, Osprey, 2013.

Kent, Percy Horace *The Passing of the Manchus*, Edward Arnold, 1912.

Keswick, Maggie *The Chinese Garden: History, Art and Architecture*, Academy Editions, 1978.

Kilpatrick, Jane *Gifts from the Garden of China,* Frances Lincoln, 2007.

Kilpatrick, Jane *Fathers of Botany, the discovery of Chinese plants by European missionaries*, Kew, 2014.

Kingdon-Ward, Frank *On the road to Tibet*, Shanghai Mercury Press, 1911.

Kissack, M Elizabeth *The Life of Thomas Hayton Mawson, Landscape Architect 1861–1933*, author, 2006.

Köll, Elisabeth *Railroads and the transformation of China,* Harvard University Press, 2019.

Lancaster, Roy *Travels in China: a Plantsman's Paradise*, Antique Collectors' Club, 1989.

Lemmon, Kenneth *Golden Age of Plant Hunters*, Dent, 1968.

Lewis, Cecil *So Long Ago, So Far Away*, Luzac Oriental, 1997.

Lovell, Julia *The Opium War*, Picador, 2011.

Lyte, Charles *Frank Kingdon-Ward: The Last of the Great Plant Hunters*, John Murray, 1989.

McLean, Brenda *A Pioneering Plantsman: A K Bulley and the Great Plant Hunters*, HMSO, 1997.

McLean, Brenda *George Forrest, Plant Hunter*, Antique Collectors' Club/Royal Botanic Garden Edinburgh, 2004.

McLewin Will & Dezhong Chen *Peony Rockii and Gansu Mudan*, Wellesley-Cambridge Press, 2006.

Menzies, Nicholas K *Forest and Land Management in Imperial China*, St Martin's Press, 1994.

Meyer, Karl and Brysac, Shareen *Tournament of Shadows: the Great Game and the Race for Empire in Asia*, Little, Brown, 1990.

Mitchell, W R *Reginald Farrer At Home in the Yorkshire Dales*, Castleberg, 2002.

Morrison, G E *An Australian in China*, Horace Cox, 1895.

Musgrave, Toby *The Plant Hunters*, Ward Lock, 1998.

Musgrave, Toby *An Empire of Plants,* 2000.

Nicolson, Juliet *The Perfect Summer: Dancing into Shadow in 1911*, John Murray, 2006.

Noltie, Henry *Indian Forester, Scottish Laird: the botanical lives of Hugh Cleghorn of Stravithie*, Royal Botanic Garden Edinburgh, 2016.

O'Brian Seamus *In the Footsteps of Augustine Henry,* Garden Art Press, 2011.

Orlean, Susan *The Orchid Thief: a true story of beauty and obsession*, Random House, 1998.

Pearl, Cyril *Morrison of Peking*, Angus & Robertson, 1967.

Porter, Bernard *The Absent-Minded Imperialists: Empire, Society and Culture in Britain*, Oxford, 2004.

Porter, Bernard *Britannia's Burden: Political Evolution of Modern Britain*, Hodder, 1990.

Presant, O V *A Perfect Friend, the life of Cumbrian plant hunter William Purdom*, Hayloft Press, 2019.

Ramsay, Robert *The languages of China*, Princeton University Press, 1987.

Reece, Bob and Cribb, Philip *Hugh Low's Sarawak Journals 1844–46*, Natural History Publications Borneo, 2002.

Reinikka, Merle A *A History of the Orchid*, University of Miami, 1992.

Richards, John *Primula*, Timber Press, 1993.

Rocco, Fiammetta *The miraculous fever tree*, Harper Collins, 2003.

Rose, Jonathan *The Intellectual life of the British Working Classes*, Yale, 2001.

Rose, Sarah *For all the tea in China*, Hutchinson, 2009.

Royle, Edward *Modern Britain: A Social History 1750–1985*, Edward Arnold, 1985.

Sargent, Charles Sprague (editor) *Plantae Wilsonianae*, 3 volumes), Cambridge University Press, 1913–17.

Seagrave, Sterling *Dragon Lady: the life and legend of the last Empress of China*, Macmillan, 1992.

Shephard, Sue *Blue Orchid and Big Tree: Plant Hunters William and Thomas Lobb*, Redcliffe Press, 2014.

Shephard, Sue *Seeds of Fortune: a gardening dynasty*, Bloomsbury, 2003.

Sheridan, James E *Chinese Warlord, the career of Feng-Yu-hsiang*, Stanford University Press, 1966.

Shulman, Nicola *A Rage for Rock Gardening: The Story of Reginald Farrer*, Short Books, 2002.

Spence, Heather *British policy and the 'development' of Tibet 1912–1933*, University of Wollongong, 1993.

Stone, Daniel *The Food Explorer* Dutton, 2018.

Stursberg, Peter *No foreign bones in China*, University of Alberta, 2001.

Sutton, Stephanne Barry *Charles Sprague Sargent and the Arnold Arboretum*, Harvard, 1970.

Thubron, Colin *Behind the Wall*, Heinemann, 1987.

Trevor-Roper, Hugh *The Hermit of Peking: the hidden life of Sir Edward Backhouse*, Alfred A. Knopf, 1977.

Valder, Peter *The Garden Plants of China*, George Weidenfeld and Nicolson, 1999.

Van de Ven, Hans *Breaking with the Past: The Maritime Customs Service and the Global Origins of Modernity in China*, Columbia, 2014.

Varè, Daniele *Laughing Diplomat*, John Murray, 1938.

Varg, Paul A *The Life of W W Rockhill* Greenwood Press, 1952.

Von Arnim, Elizabeth *Elizabeth and her German Garden*, Macmillan, 1898.

Von Arnim, Elizabeth *The Enchanted April*, Macmillan, 1922.

Watkins, Charles *Trees, Woods and Forests A Social and Cultural History*, Reaktion Books, 2014.

Watt, Alistair *Purdom and Farrer, Plant Hunters on the Eaves of China*, privately printed, 2019.

Wharton, Peter (editor) *The Jade Garden: New and Notable Plants from Asia*, Timber Press, 2005.

Wilson, Ernest *A Naturalist in Western China*, Methuen, 1913.

Wilson, Ernest *Aristocrats of the Trees*, Stratford, 1930.

Winstanley, David (translator and editor) *A Botanical Pioneer in South West China: Heinrich Handel-Mazzetti*, Winstanley, 1996.

Wood, Frances *Betrayed Ally: China in the Great War*, Pen & Sword, 2016.

Wynne-Thomas, Peter *Ivo Bligh*, Association of Cricket Historians, 2002.

Zhao Changtian *An Irishman in China: Robert Hart, Inspector General of the Imperial Maritime Customs*, Better Link Press, 2014.

Articles, periodicals and academic theses

Anon Sketch of Nicholas Prejevalski, *Popular Science* 1887, 30.

Anon *Second Exhibition Supplement*, the Royal International Horticultural Exhibition, 1912.

Anon Primula Conference, *Journal of the Royal Horticultural Society* 1913, Vol 39, pp. 98–161.

Bailie, Joseph Magna Charta of China's Forestry Work, *American Forestry* 1916, Vol 22, pp. 269–272.

Curtis, C H The House of Veitch (article in two parts), *Journal of the Royal Horticultural Society* 1948, Vol LXXIII, Part 8 and Part 9.

Dosmann, Michael S A Lily from the Valley, *Arnoldia*, Vol 77 no. 3 pp 14–25.

Ejder, Ridderlöf and Salomonsson Rhododendron purdomii of the Taibai Shan *2021 Yearbook of the Magnolia, Camellia and Rhododendron Group, Royal Horticultural Society*.

Farrer, R F Series of articles about the 1914/15 expedition, *Gardeners' Chronicle*: 1914, 56, pp. 185, 213, 258–259, 287, 318–319; 1915, 57, pp. 1, 109–110, 143, 193–194, 217, 231–232, 257–258, 289–90, 325, 337–338; 1916; 59, pp. 17, 29–30, 45–46, 59–60, 72–73, 86–86, 100–101, 127–128, 139–140.

Farrer R F Report of work in 1914 and 1915 in Kansu and Tibet, *Journal of the RHS* 1916–17, Vol XLII, pp. 47–52 and 324–348.

Gilliard D, *Education in England, a history* (2018), www.educationengland.org.uk/history

Harrow, George Some recollections of Coombe Wood, *The New Flora and Silva* 1931.

Heriz-Smith, Shirley Brightener of British Winters, William Purdom in China and Tibet, *Country Life* 1986, June 5.

Heriz-Smith, Shirley Japan rooted in Surry, *Country Life* 1992, April 23, pp. 49–51.

Heriz-Smith, Shirley Veitch & Sons of Chelsea and Robert Veitch & Son of Exeter, 1880–1969, *Garden History* 1993, Vol 21, 1, pp. 91–109.

Heriz-Smith, Shirley William Purdom (1880–1921), A Westmorland Plant-hunter in China, *Hortus* 1996, Vol 38, pp. 49–62.

Hers, Joseph Notes sur les saules et peupliers de la Chine du Nord, *Bulletin de la Société Dendrologique de France* 1923, Vol 49, pp. 152–159.

Hers, Joseph Notes sur les conifères de la Chine du Nord, *Bulletin de la Société Dendrologique de France* 1923, Vol 49, pp. 160–171.

Hers, Joseph Les forêts du nord-est de l'Asie, *Bulletin de la Société Dendrologique de France*, 1926, Vol 59, pp. 59–65.

Holway, Tatiana M History or Romance? Ernest H Wilson and plant collecting in China, *Garden History* 2018, Vol 46, 1, pp. 3–26.

Kelley, Susan Plant Hunting on the Rooftop of the World, *Arnoldia* 2001, 61/2.

King, Pan Chen Suggestions for Forest Administration in China, *Forestry Quarterly* 1914, pp. 578–592.

Lancaster, Roy William Purdom, *The Lakeland Gardener* Spring 1981.

Lewis, James G Theodore Roosevelt's Cautionary tale, *Forest History Today*, Spring/Fall 2005. (This article includes the full text of Theodore Roosevelt's 1908 address to Congress).

Lu Wei-zhong, Ren Ji-wen Plant biodiversity and its conservation in Maijishan Scenice Regions of Gansu, *Journal of Northern Forest University*, 2005–04.

Mather, Jeff Camping in China with the Divine Jane, *Journeys* 2010, Vol 10, 2.

Meyer Frank N Economical Botanical Explorations in China, *Transactions of the Massachusetts Horticultural Society* 1916, pp. 125–130.

Morell, Virginia The mother of Gardens, *Discover Magazine* 6 August 2005.

Mueggler, Eric The Lapponicum Sea: Matter, Sense and Affect in the Botanical Exploration of South West China and Tibet, *Comparative Studies in Society & History* 2005, Vol 47, 3, pp. 442–449.

Noblett, H William Purdom (1880–1921) Plant Collector from Westmorland, *The Lakeland Gardener* Spring 1975.

Pearson, Lisa Picturing China, The expeditions and photographs of William Purdom, *Silva* Fall-Winter 2015–2016.

Pitts, Larissa Noelle *Seeing the Forest from the Trees, Scientific Forestry and the Rise of Modern Chinese Environmentalism*, PhD thesis, UCL, Berkeley, 2017.

Setzekorn, Eric The First China Watchers: British Intelligence Officers in China, 1878–1900, *Intelligence and National Security* 2013, Vol 28, 2, pp. 181–201.

Sherfesee, F The Reforestation Movement in China, *American Forestry* 1915, vol XXI, pp. 1033–1040.

Sherfesee, F China's Forest Laws, *Forestry Quarterly* 1916, pp. 650–661.

Sherfesee, F The Industrial and Social Importance of Forestry in China, *The Chinese Social and Political Service Review* 1916, Vol 1, 3, pp. 1–26.

Sherfesee, F Organisations and Activities of the Chinese Forest Service, *Transactions of the Royal Scottish Arboricultural Society* 1917, Vol XXXI, pp. 115–130.

Songster, Elena E Cultivating the nation in Fujian's forests, forest policies and afforestation efforts in China 1911–1937, *Environmental History* July 2003.

Toshio Yoshida, Hang Sun *Meconopsis Lepida* and *M. Psilonomma (Papaveraceae)* rediscovered and revised, *Harvard Papers in Botany* 2017, Vol 22, 2, pp. 157–192.

Wagner, Jeff From Gansu to Kolding: the expedition of J F Rock in 1925–27 and the plants raised by Aksel Olsen, *Annual Journal of the Danish Dendrological Society* 1992, Vol X, pp. 19–93.

Wang Jiyun, Che Kejun, Yan Wende Analysis of the Biodiversity in Qilian Mountains, *Journal of Gansu Forestry Science and Technology*, 1996–02.

Zuo Cheng Ying The botanical exploration of William Purdom in China in the early 20th Century, *Henan University Journal of History and Science, [Shiksue Yuekan]* 2020 issue 8, pp. 40-58.

Chinese archival sources

My collaborator, Cheng Yi Meng, searched the following Chinese archives for me

The Second Historical Archives of China, Nanjing

Ministry of Forestry and Agriculture archives (which contain the exchange of diplomatic notes concerning Will's arrears of salary)

Academia Sinica

He also searched, in vain, the following

Chinese Agricultural University Library

Henan Provincial Archives

Jiangou Provincial Archives

Xinyang Provincial Archives

BIBLIOGRAPHY

Professor Wang Huoran of the Chinese Academy of Forestry kindly provided some material concerning Han An, and advised that the Academy archives hold no papers relating to Will Purdom.

The history of China during the decades following Will's death records is deeply troubled, and many important records have been destroyed. I suspect, however, that there is more to be discovered concerning Will in Chinese archives, for example in records relating to Republican-era railway companies, which are currently closed to foreigners.

Online sources

More and more archives are digitising their collections and making them available online, greatly to the benefit of researchers. Logically, the next step is the creation of digital archives concerning discrete topics and combining in one virtual space the material held by different institutions.

Leyla Cabugos from the University of California David Library is the convenor of a group comprising herself, Lisa Pearson from the Arnold Arboretum, and Pamela Nett Kruger from California State University Library, which holds the US Department of Agriculture records from the Chico Plant Introduction Garden. So far, the group has placed online all four volumes of Meyer's correspondence, and has augmented the Wikipedia entry for Meyer with more detailed biographical material and inserted links to archives and herbarium collections. The work of the group is ongoing and it is proposed to catalogue in greater detail more of the material held by the different institutions, and eventually to digitise it.

Whether or not the above yet constitutes a digital Frank N Meyer archive is a question of definition, but it certainly makes it very much easier to research Meyer's life and work.

At time of writing, the best entry-point is via the UCD catalogue at https://search.library.ucdavis.edu/permalink/f/u9eqg/01UCD_ALMA51328468250003126

which contains a link to all the documents relating to Meyer in the Internet Archive. Meyer deserves no less, and this is an exciting precedent.

Malcolm Peaker FRS, FRSE, has researched in depth the 1904–11 zoological expedition to north-west China and Tibet sponsored by the Duke of Bedford and has posted four detailed articles about the expedition and the persons involved on zoologyweblog.blogspot.com.

Endnotes

[1] William Dallimore, *A Gardener's Reminiscences*, unpublished manuscript c. 1950 held in the archives of the Kew Botanic Gardens, pp. 501–502.

[2] Tom Steele, *Knowledge is Power,* OUP 2007, p. 54.

[3] James Abbott letter of application to Kew 15 May 1894 contained in his personnel file, Kew Royal Gardens archive.

[4] This account of the Kew student gardeners scheme draws heavily on W A Lord's article in *The Journal of the Kew Guild* [JKG] 1979, pp. 780–791.

[5] Fiametta Rocco, *The Miraculous Fever Tree*, Harper Collins, 2003.

[6] Joe Jackson, *The Thief at the End of the World*, Duckworth Overlook 2008.

[7] *New York Times,* 22 September 1895.

[8] James Abbott graduated from Kew in 1898 and set up on his own account as a nurseryman in Rushden, Northamptonshire: see *JKG* 1899–1943 for details.

[9] *Gardeners' Chronicle*, Dec 3, 1921, p. 294.

[10] Reginald Farrer [RJF] letter to Aubrey Herbert, 7 September 1902, DD/HER 38, Somerset Archives, Taunton.

[11] Jim Endersby, *Orchid a cultural history,* Kew Publishing 2016, citing *Strand Magazine* 1906, 81, p. 31.

[12] Will Purdom (WP) Personal File (PF), Kew, James Abbott letter 14 June 1902.

[13] Merle A Reinikka, *A history of the orchid,* Timber Press, Oregon, 1995, p. 63.

[14] Reece and Cribb *Hugh Low's Sarawak Journals*, Natural History Publications (Borneo) 2002, p. 24.

[15] Arnold Arboretum [AA] archive, Harry Veitch [HV] letter to Charles Sprague Sargent [CSS], 10 January 1893.

[16] Charles Curtis, *op. cit.,* p. 247.

[17] A H Unwin, *West African Forests and Forestry*, London, 1920), cited in Fairhead and Leach *Dessication and Domination*, Journal of African History Vol 41, no 1.

[18] James G Lewis, *Forest History Today,* Spring/Fall 2005, pp. 53–57. See also the complete text of Theodore Roosevelt's eight annual message to Congress, 8 December 1908, at presidency.ucsb.edu/documents/eighth-annual-message.

[19] Ray Desmond, Kew, the history of the Royal Botanic Gardens, RBG Kew, 1995.

[20] Dallimore, *op cit,* p. 483.

[21] Dallimore, p. 440.

[22] Dallimore, p. 464.

[23] Board of Agriculture and Fisheries file MAF 46/50, Public Record Office (PRO), Kew.

[24] Dallimore p. 498.

[25] Dallimore, p. 503.

[26] Dallimore, p. 504.

[27] The principle that the public service should pay *"such wages as are generally accepted as current in each trade for competent workmen"* had been unanimously approved by the House of Commons in 1891.

[28] Robert Ensor, *England 1870–1914*, OUP 1936, pp. 115–117.

[29] Ensor, *op cit*, p.118.

[30] WP letter 17 July 1905 and LRC reply 18 July 1905, docs LP.GC 6/278 and 7/278, Labour History Archive, Peoples Museum of History, Manchester.

[31] *JKG* 1933, p. 263.

[32] Dallimore, p. 508.

[33] CSS letter to David Prain [DP] 21 Dec 1905, AA archive.

[34] Hansard 25 Feb 1908, Vol 184 columns 1579–80.

[35] Robert C Allen, *Real Incomes in the English-speaking world*, in Labour Market Evolution, Routledge, 1994.

ENDNOTES

36 DP minute PRO 1138/1908.

37 Dallimore p. 531.

38 Letters of 22 and 24 April 1909, documents LP.GC14/252 and LP.GC 14/253, Peoples Museum of History, Manchester.

39 Dallimore, p. 531.

40 Frank Meyer letter to Ernest Wilson, May 7, 1907, AA archive.

41 HV letter to CSS 14 January 1909, AA archive.

42 Note in Purdom family archive.

43 Ernest Wilson [EHW] letter to CSS, 24 Nov 1908, Arnold Arboretum archive.

44 JKG 1906, p. 294.

45 Tatiana Holway, *History or Romance?* Garden History, the Journal of the Gardens Trust, Vol 46:1 pp. 3–27.

46 WP letter to CSS, 26 March 1909, AA archive.

47 WW Rockill, *The 1910 census of the population of China*, Leyden, 1912, pp. 117–125.

48 WP letter to CSS, 1 April 1909, AA archive.

49 Nell Purdom letter to Lakeland Horticultural Society [Lakeland Horticultural Society], 9 Feb 1977, Lakeland Horticultural Society Archive.

50 Nell Purdom letter 9 Feb 1977.

51 *Sweet Stone Bridge*, p. 29.

52 WP letter to HV, 27 April 1909 and letter to CSS 2 May 1909, both AA archive.

53 Nell Purdom letter 9 Feb 1977.

54 Jack Garnett (Third Secretary, British Legation, Peking) letter to his mother, March 23 1906, Lancashire Record Office, Preston, Garnett Collection, doc DDQ 9/12/21.

55 A J Farrington, *British military Intelligence on China and the Boxer Rising, c.1880-1930*, Leyden 1904, pp. 1–3.

56 Han Suyin, *The Crippled Tree*, London 1965.

57 Wu Ting-Fan, *The Causes of the Unpopularity of the Foreigner in China*, Annals of the American Academy of Political and Social Science, Vol 17 (January 1901), p. 8.

58 See also Wu Ting-Fan, op. cit. p. 9.

59 WP letter to CSS, 11 May 1909, AA archive.

60 WP letter to CSS, 18 May 1909, AA archive.

61 Letter in the AA archive.

62 WP letter to CSS, 18 June 1909, AA archive.

63 WP letter to CSS, 5 July 1909, AA archive.

64 WP letter to CSS, 26 August 1909, AA archive.

65 WP letter to HV, 5 Sep 1909, AA archive.

66 WP to CSS, 6 Oct 1909, AA archive.

67 WP letter to HV, 22 Dec 1909, AA archive.

68 WP letter to CSS, 6 Oct 1909, AA archive.

69 CSS letter to WP, Feb 8 1909, AA archive.

70 CSS letter to WP, 8 Nov 1908, AA archive.

71 G. E. Morrison diary entry 14 January 1910, George E. Morrison Collection, Mitchell Library, New South Wales State Library.

72 Susan Kelley *Plant Hunting of the Rooftop of the World*, Arnoldia 61/2 (2001). See also Virginia Morell *The Mother of Gardens*, Discover Magazine, 6 August 2005.

73 Wang Jinye, Che Kejun, Yan Wende, *Analysis of the Biodiversity in Qilian Mountains*, Journal of Gansu Forestry Science and Technology 1996–02, also Lu Wei-Zhong, Ren Ji-wen *Plant Biodiversity and its Conservation in Maijishian Scenic Regions of Gansu*, Journal of Northwestern Forestry University 2005-2.

74 Lancaster Record Office, Preston, Garnett collection, DDO/21/4.

75 HV letter to CSS, 25 March 1911, AA archive.

76 WP letter to CSS, 24 June 1911, AA archive.

77 Howard Van Dyck, *William Christie, Apostle to Tibet*, Christian Publications Inc. 1956, p. 67.

78 Note in WP personal photo album, RJF 3/3, Royal Botanic Garden Edinburgh archive.

79 S B Sutton, *In China's Border Provinces: the turbulent career of Joseph Rock*, Hastings House 1974, p. 153.

80 Reginald Farrer, *On the Eaves of the World*, Edward Arnold, London 1917, p. 305.

81 Jonathan Fenby, *The Penguin History of Modern China,* Penguin 2008, p. 115.

82 Sir John Jordan, despatch to Foreign Office, 15 January 1912, PRO file FO 371/1310, folio 554.

83 Harold Frank Wallace, *The Big Game of Central and Western China*, John Murray 1913 pp. 183 *et seq.*

84 Nell Purdom letter to Lakeland Horticultural Society, 9 February 1977, Lakeland Horticultural Society archive.

85 WP letter to HV, 23 March 1912, AA archive.

86 *Daily Mail* July 10 1913, p. 3. The reporter cited the alleged use of a revolver. Will did possess a pistol, but he also owned a lever-action rifle, a far more effective weapon.

87 WP letter to HV, 23 March 1912, AA archive.

88 Note in Will's handwriting on the back of a photograph in the AA archive.

89 WP letter to Reginald Farrer, 9 September 1915, Royal Botanic Garden Edinburgh archive.

90 HV letter to CSS, June 9 1911, AA archive.

91 *Plantae Wilsonianae,* Charles Sprague Sargent (editor), Arnold Arboretum publications, 1913, Vol. I, p. 500.

92 RHS Lindley Library PUR/2/1.

93 CSS letter to EHW, 9 February 26 1901, AA archive.

94 Sue Shephard, *Seeds of Fortune*, 2003, p. 259.

95 HV letter to CSS 15 November 1911, also letter of 31 January 1912, both in the AA archive.

96 Journal of the Kew Guild, 1913, p. 88.

97 FN Meyer [FNM] letter to David Fairchild October 15 1912, USDA compilation of Fairchild correspondence held at Peter J Shields Library, University of California, Davis, Vol 3, pp. 1600–1601.

98 FNM letter to David Fairchild, November 14 1912 USDA compilation Vol 3, pp. 1619–1621.

99 FNM letter to Fairchild December 21 1912, USDA compilation, Vol 3, p. 1630.

100 FNM letter to Fairchild, November 14 1912 USDA compilation Vol 3, pp. 1619–1621.

101 Charles H Hough*, A Westmorland Rock Garden*, privately printed 1929, p. 18.

102 *The Lake District Herald*, November 26 1921, "A tribute to the late William Purdom junior", cutting in William Purdom's scrapbook/photo album, Purdom family papers.

103 Morrison collection, Mitchell Library, State library of New South Wales, Vol 75 pp. 137–141.

104 GEM letter to WP, 5 August 1913, Morrison collection, Vol 75, p. 277.

105 *Gardeners' Chronicle*, October 4 1913, pp. 229–231.

106 Journal of the RHS, Vol 39 (1913) pp. 98–195.

107 Charles H Hough, *op cit.* p. 18.

108 WP letter to RJF, 9 September 1913, Reginald Farrer collection, RJF/2/1/4, Royal Botanical Garden Edinburgh archives.

109 Osbert Sitwell, *Noble Essences*, Macmillan, 1950, p. 17.

110 This measurement is derived from the analysis of a single photograph undertaken by an expert retired RAF officer, which involved several pages of trigonometric formulae.

111 EHM Cox, *Farrer's last journey*, Dulau & Co 1926, p. 2.

112 On the Eaves of the World, p. 1.

113 The text of the will can be found (for a fee) at www.gov.uk/search-will-probate.

114 WP letter to Morrison, 26 March 1914, Morrison Collection, Vol.79, pp. 185–187.

115 *On the Eaves of the World*, pp. 167–169

ENDNOTES

[116] Toshio Yoshida and Hang Sun, *Meconopsis Lepida and M. Psilonomma rediscovered and revised*, Harvard Papers in Botany Vol 22 (2017) pp. 157–192.

[117] F N Meyer letter to Fairchild, December 10 1914, USDA compilation Vol 3, pp. 1986–1992.

[118] On the Eaves of the World, p. 305.

[119] On the Eaves of the World, p. 309–310.

[120] *Gardeners' Chronicle*, October 3, 1914, p. 241.

[121] *Gardeners' Chronicle*, February 27, 1915, p. 109.

[122] Marjorie Acheson, transcript of 8 Dec 1980 interview, Ambleside Oral History Group collection, Ambleside Public Library.

[123] Anne Wilkinson, *The Development of Gardening as a leisure activity and the establishment of Horticultural Periodicals*, PhD thesis, the Open University 2002, pp. 106–110 and 131.

[124] George Fenwick-Owen and Walter Fenwick were second cousins once removed: see family trees 1, 3, and 7 in *Fenwicks of the North* (2017) by Roger Mott.

[125] Quoted in the Lake District Herald Nov 26 1921, *A tribute to the late William Purdom junior*.

[126] G E Morrison diary 30 December 1915, Mitchell library.

[127] Caption to photo at p. 46 Farrer photo album RJF 2/2/4, Royal Botanic Garden Edinburgh archive.

[128] Shulman *A Rage for Rock Gardening*, Short Books, 2002, p. 105.

[129] cf. RJF letters to Lady Celia Brunel Noble, 1920, Farrer collection, Royal Botanic Garden Edinburgh archive.

[130] RJF letter to Isaac Bayley Balfour, Dec 13, 1915, Farrer collection, Royal Botanic Garden Edinburgh archive.

[131] *Peking Gazette* Jan 15 1916.

[132] Morrison diary 30 Dec 1915, Mitchell Library, also Frank N Meyer letter to Stephen Stuntz, USDA, June 23 1917, pp. 2343–2348 Meyer correspondence.

[133] Morrison diary 11 January 1916, Mitchell Library

[134] WP notebook, PUR 1/1/, RHS archive, Lindley Library.

[135] RJF letter to Morrison 17 April 1916, Morrison collection, Mitchell Library, Vol. 87 pp. 415–418.

[136] Morrison letter to RJF, 19 April 1916, Morrison collection, Mitchell Library, Vol. 87 pp. 453–457.

[137] Morrison letter to WP 19 April 1916, Morrison collection, Mitchell Library, Vol. 87 pp. 459–463.

[138] RJF letter to Morrison, 14 May 1916, Morrison Collection, Mitchell Library, Vol 88 pp. 51–52.

[139] Confusingly, Morrison kept *two* diaries during at least part of 1915, both of which are in the Mitchell library. One was a pocket diary printed in China which also served as an appointments book and one a much larger book which was clearly kept with an eye to eventual publication or (Morrison was not a modest man) for the benefit of future historians. The account of Will's *"funny stories"* comes from the pocket diary for 18 December and was not copied by Morrison into the larger diary. Unless otherwise specified, references to Morrison's diary are to the latter.

[140] Newspaper article, *Trees and Chinese Welfare*, Peking Gazette, Oct 17, 1916.

[141] RJF letter to Jane Purdom, 5 November 1916, Purdom family papers.

[142] WP letter to Farrer, 1 Dec 1916, Royal Botanic Garden Edinburgh archive.

[143] Brenda McLean, *George Forrest, Plant Hunter*, ACC/Royal Botanic Garden Edinburgh 2004, p. 131.

[144] RJF letter to Mrs E Farrer, 22 February 1918, Farrer Collection, Royal Botanic Garden Edinburgh archive.

[145] Frances Wood and Christopher Arnander, *Betrayed Ally: China in the Great War*, Pen & Sword 2016, pp. 71–9.

[146] Document PUR/1/2, RHS Lindley Library archive.

[147] RJF letter of 17 May 1917 to E A Bowles, RHS archive, doc EAB 2/3/5.

[148] *Lake District Herald* tribute 21 November 1921.

[149] Hsi-Sheng Ch'i, *Warlord Politics in China 1916-1928*, Stanford University Press, 1976.

[150] G E Morrison letter to Sir John Jordan, 6 May 1920, Vol. 2 edited correspondence, pp. 816–7.

[151] Boorman, *Biographical Dictionary of Republican China*, Columbia University Press, 1970, volume 3, p. 367.

[152] Köll, Railroads and the Transformation of China, Harvard, 2019, pp. 61–61.

[153] Köll, *op cit, pp.* 67–68.

[154] John H Reisner, *Progress of Forestry in China in 1919–1920*, Journal of Forestry, Vol 19 Issue 4 (April 1921) p. 396.

[155] William Judd diaries (unpublished manuscript), Vol. 2, p 261, AA archives IV A-1 W.

[156] Hers, Joseph, *Bulletin de la Société Dendrologique de France*, 59 (1926) p. 59.

[157] Kew Inward Register, 1916-27, pp. 99, 123, 139.

[158] The compliment slip is inserted into the Veitch ledger in the RHS Lindley Library.

[159] Reisner op.cit.

[160] WP letter to RJF, 10 June 1920, Royal Botanic Garden Edinburgh archive.

[161] This chapter draws heavily on the private collection of papers held by the Leveson family to which they generously gave me unrestricted access.

[162] Bodley, *The Coronation of Edward the Seventh*, Methuen, 1903, pp. 335–345.

[163] Betty's novel about life in Peking, *Legation Street*, which is written in the first person, includes extended musings by the narrator about the difficulties encountered by married women trying to retain their independence and individual identity.

[164] Letter from Molly Orpen-Palmer, wife of the British Military Attaché at the British Legation in Peking, to her mother, 18 January 1921, private collection.

[165] *The Peking Daily News* 22 February 1921, cutting in Leveson family papers.

[166] WP letter to RJF April 11 1917, Royal Botanic Garden Edinburgh archive.

[167] Wynne-Thomas & Griffiths, *Ivo Bligh*, Association of Cricket Historians, 2002, pp. 7–8.

[168] Baroness Clifton, *Living in the Country*, Country Life Ltd, 1934, p. 189.

[169] *Living in the Country*, p. 188.

[170] *Journal of the Kew Guild,* 1921 (December), p. 115.

[171] The death certificate issued by HM Consul did not give the cause of death, which was ascribed by the Chinese press to rheumatic fever and pneumonia. The retired senior consultant physician who reviewed the sparse case history for me concluded that it supports the hypothesis of an infection which in this pre-antibiotic era became overwhelming. Another possibility is that the lumps in Will's neck were the manifestation of a malignant process which caused what would admittedly have been a rather rapid post-operative decline.

[172] Undated cutting in William Purdom's cuttings album, Purdom family collection.

[173] William Purdom cuttings and photo album, Purdom family collection.

[174] *Gardeners' Chronicle*, Dec 13 1921, p. 294.

[175] *Journal of the Arnold Arboretum*, Vol 3, 1921, pp. 55–56.

[176] Charles Hough, *A Westmorland Rock Garden*, Frederic Middleton, 1929.

[177] Hers, Joseph, Notes sur les conifères de la Chine du Nord, *Bulletin de la Société Dendrologique de France* 49 (1923) p. 170.

[178] Boorman, *op. cit.,* p. 369

[179] The tree is on the Orangery (north) side of the White Peaks café, in the north-western part of Kew Gardens. It bears a label with its botanical name.

Index

References to illustrations are indicated in **bold**. References to footnotes include the appropriate symbols as required. References to endnotes consist of the page number followed by the letter 'n' followed by the number of the note with the page number from the main text in brackets, e.g. 228n56 (67).

Abbott, James 5–6, 8, 17
abelias 84, 111
acers
  Acer hersii 181
  Acer grosseri (formerly *Acer hersii*) 181§
Acho-che-ro (Choni resident) 98, **99**
aconites 111
Aesculus chinensis 89, **106**, 107–8, 111, 112, 121, 122, 202, **212**
'The afforestation question in China' (Purdom's memorandum) 108–9, 110, 213–14
Africa, *Eucalyptus grandis*, propagated by Kew 20
Ajiao mountains 147, 160
Alhagi maurorum (camel-thorn) 20
alliums 183
alpines 127, 128, 131, 137, 159
Alston, Beilby 186–7
Ambleside
  Free Grammar School 4
  Purdom's obituary in parish magazine 195
  see also Brathay Hall (Ambleside)
"ambush" incident (March 1912) 106, 107, 120
American Legation, Peking 62
  see also United States
American swamp cypress (*Taxodium distichum*) 180–1, **209**
Ames, Oakes, Prof. 200
Anderson, John Abbott, Colonel 63, 64–5, 68, 74, 77–8, 87–8

androsaces (rock jasmine) 150
anemones 68, 70–1, 113, 139
animals *see* wildlife
Annam kingdoms (now Vietnam), France's control over 55
Aracauria (monkey puzzle tree) 13
Arnim, Elizabeth von 198
Arnold, James 39, 115
Arnold, William 12*
Arnold Arboretum, Boston
  archives i
  Chinese Sargent–Veitch expedition (1909–12) 43, 48–9
    plant material sent by Purdom 112, 113, 117
  Director's *Annual Report to President of Harvard, 1910–11*, Wilson's and Purdom's results compared 114–15
  Hengduan plant-collection expedition (1997) 79, 81
  Hers' herbarium specimens 181–2, 199
  Plantae Wilsonianae 85*, 92, 112
  plants
    Aesculus chinensis, reintroduction of 107–8, 112
    Prunus Padus var. *pubescens* forma *Purdomii* 70*
    Rhododendron purdomii herbarium specimen (1910) 88, 196, **206**
  Purdom, number of plants introduced and collected by 196
  Purdom, plant material from (1920–21) 182–3, 184
  Purdom's death, notice of 196
  Purdom's rejection of Assistant Superintendent post 39–41
  Rock's wild peony expedition (1924) 92†

  Sargent as Director 14, 39, 75
  Sargent's death and new Director 200
  Sargent's fundraising for 115
  Sargent's interest in Moutan (tree) peonies for 91
  Wilson's Chinese expedition (1906–07) 43
  Wilson's post as Keeper 200
  Wilson's post (taxonomical work) 78
  Wilson's recorded specimens (1910–11) 113
  see also Sargent, Charles Sprague
Asquith, Raymond 170
Aster flaccidus var. *purdomii* 125
Australia
  camel-thorn (*Alhagi maurorum*), propagated by Kew 20
  Eucalytus species, translocation of to Kenya and Uganda 7
  libraries for sources on Purdom i
  see also Mitchell Library (State Library of New South Wales)

Backhouse, Edmund, 2nd Baronet 58*
Bai Lang ('White Wolf') 134, 137, 139, 140
Balfour, Arthur 33
Balfour, Isaac Bayley 127–8, 133, 152–3, 161
bandits *see* highwaymen
Baoding *see* Pao-ting-fu, Shansi (now Baoding, Hebei Province)
Barton (family) 188
Barton, Esmé 187
Barton, Mary (Mrs) 186, 187, 189
Barton, Sidney 186–7
Bean, William 44
Bedford, Hastings Russell, 12th Duke of 93, 102

INDEX

Bedgebury Pinetum 21
Bee's Nursery 152, 160
Belgium
  National Botanical Gardens 182, 199
  treaties with China (19th century) 54
Bentham, Jeremy 19
berberis 139
Betula utilis ssp albosinensis (red-barked birch, formerly *Betula albosinensis*) 91, 126
biopiracy 7*
birches 10, 48, 68, 183
  red-barked birch (*Betula albosinensis*) 91, 126
Bligh, Edward *see* Darnley, Edward Bligh, 7th Earl of Darnley and Baron Clifton
Bligh, Ivo *see* Darnley, Ivo Bligh, 8th Earl of Darnley
Board of Agriculture and Fisheries (UK) 26, 27, 33, 34, 35, 36–7, 38, 42
botany
  economic botany 20, 22
  Index Kewensis 23
  scientific botany 22
  see also plants
Bourne, Thomas 172–3
Bowles, Augustus 114*, 174
Boxer Rebellion (China, 1899–1901) 53, 55, 56, 57, 58
Boyer, Mr and Mrs 149
Brathay Hall (Ambleside)
  Brathay estate
    conservatory 3, 10
    gardens 3, 9
    poor soil 9
    trees 3, 9–10
  Brathay Lodge (head gardener's cottage) **2**, 3, 47
    Hall from Lake Windermere 1
  James Abbott's apprenticeship 5–6, 8, 17
  Will Purdom's apprenticeship 8, 9, 10
  William Purdom as head gardener 3, 5, 10
Bretschneider, Emil, *European Botanical Discoveries in China* 83
Britain *see* United Kingdom
British Army in India, China Intelligence Section 64–5
British Empire
  Hugh Low's botanical/horticultural activities, profit from 13
  Kew's links with botanical gardens across Empire 6–8, 20
British Gardeners' Association 25, 26, 28, 29
British Isles, number of plant species 79*
British Legation, Peking
  advice not to travel to western China (1914) 134
  Anderson, attached to 63, 64
  British secret services, contacts with 65, 117, 175–6
  Clifton's *The Moon Flower*, in theatre of 187
  Hosie and commercial department of 59
  Pearson in charge of security at 133
  Prunus triloba "Multiplex" in garden of 183
  Purdom's death and dealings with Chinese authorities 197
  Purdom's first meeting with Han An 109
  Purdom's first visits on arrival 61, 62
  Purdom's memorial in chapel of 191, 202, **211**
British Museum of Natural History *see* Natural History Museum, London
British secret services 64–5, 117, 175–6, 197*
  British Secret Bureau (later Secret Intelligence Service, aka MI6) 64
  Purdom's support for Anderson's *sub rosa* activities 88
British-American Tobacco Company 177
Brooke, James 12
Brooke, John Weston 96†, 138
'brush conversation' 145
Buchan, John 172†
buddleias
  B. alternifolia 136, 142, 150–1
  B. purdomi (synonymn *B. brachystachya*) 136
Budorcas taxicolor (takin) 102
Bulley, Arthur K. 152, 160, 174
Burke, William, 'burking' 150†
Burma
  Britain's control over 55
  Farrer's death in 184
Bussière, Jean, Dr 193

Cal-ceen-Wong, Mongolia 70, **71**
Calcutta, India, Royal Botanic Gardens 20, 33
camel-thorn (*Alhagi maurorum*) 20
cameras
  Kodak 'pocket' camera 46, 73†
  Sanderson camera ('Tropical' model) 46, 66*, 73, 76, 134, **205**
  see also photographs (taken by Purdom)
Campbell-Bannerman, Henry 33
Canton, battle of (1857–58) 53
Cao Rulin *see* Tsao Ju-lin (Cao Rulin)
Caragana 71
Carrington, Charles, Lord 33, 34, 35, 38*, 42
Chago monastery 137, 138, 139, 142
Chang Chien (Zhang Jian) 163
Changtingssu monastery (nr Choni) 95, 102
Chengde *see* Jehol (modern Chengde)
chestnuts 68
Chihli province (modern Zhili)
  reforestation scheme 168
  see also Weichang district, Chihli (Zhili) province
China
  19th century uprisings 51, 53, 55
    Boxer Rebellion (1899–1901) 53, 55, 56, 57, 58
    Dungan (Muslim) Rebellion (1862–77) 53, 54
    Nian Revolt (1851–68) 53, 54
    Taiping Rebellion (1850–64) 53, 54
  anti-foreigner sentiment 55, 56, 58

INDEX

archives i, v
Canton, battle of (1857–58) 53
Confucian ethical/
social code 51–2
Cultural Revolution
(1966–76) 202
dynasties
  Ch'ing (Manchu)
    45, 51–2, 53–4, 55,
    56, 61, 67, 96, 100–1
  Ming 51, 53
  Tang 145
'Eight Power Alliance' victory
and settlement (1900–01) 56
Emperors
  Kwang-Hsu (Guangxu)
    54, 57, 62
  Pu Yi 57, 104
  Tung-chih (Tongxi)
    51*, 53–4
  Xianfeng 53, 54
Empress Dowager Cixi
(Tsu-Hsi)
  53–4, 56, 57, 58, 58*
foot-binding 57
foreign states, policy towards
(until late 18th century) 52
foreign states,
subservience to 54–5
gentry 51, 52, 55, 57
Imperial edicts (1906, 1908)
on constitutional reforms 57
Imperial Maritime
Customs Service 45, 46*
international loan (1912) 125
local militias/armies,
proliferation of 55
maps 63, 65, 117
May the Fourth Movement
(1919) 180
modernisation 56–7
Old Summer Palace,
burning of (1860) 53
opium trade/wars
  52–3, 57, 85, 103
  British-Chinese opium
    agreement (1909) 85
population growth (by 1900)
  55
Regent (Prince Chun)
  57, 103*, 104
Republican (Xinhai)
Revolution (1911)
  100–2, 103–6, 116, 125–6

Second Revolution (1913) 126
Shadong 'concession',
loss of to Japan 180
travelling in rural China
  63–4
treaty breaches, reparations
and financial problems 54–5
treaty ports 54, 56, 67
warlord era (1910s–20s) 176
warlords 54
Western merchants' trade 52
Xi'an massacre (1911)
  100, 101–2, 103
see also Chinese 1909–12
Sargent–Veitch plant-
collecting expedition;
Chinese 1914–15 Farrer
plant-collecting expedition;
Chinese languages;
Forestry Adviser to
Chinese Government
China Inland Missionary
Society 47
Chinese 1909–12 Sargent–Veitch
plant-collecting expedition
  before departure
    acceptance of position
      44–5, 46
    briefing from Augustine
      Henry 45–6, 58
    briefing from Robert Hart
      45, 46, 58
    goodbye to parents
      and fiancée 46–7
    knowledge of Chinese
      history/society 58
    learning Mandarin 46
    learning to prepare
      herbarium specimens 49
    learning to use cameras 46
    legal contract 48–9
    letters of introduction
      47, 59, 61, 62, 63
    sailing to Boston on
      Oceanic 47
    sailing to Shanghai on
      Empress of Japan 49
    salary per annum
      44, 58†, 128‡
    Sargent's memorandum
      of 'guidance' 47–8
    Veitch's goals 48
  map of Purdom's movements
  (1909–12) **72**

year 1909 (arrival and Peking)
  arrival in Peking 61–2
  arrival in Shanghai
    49, 51, 58, 61, 67
  briefing from Ernest Wilson
    47, 58–9
  Chinese name (Pao Er Deng)
    63
  exploring Peking 68
  improved economic/
    social status 65–6
  learning Mandarin 62–3
  lodgings with Legation staff
    62
  meeting with
    Alexander Hosie
      59, 61, 62, 66
  meeting with
    William Rockhill 62
  passport formalities 62, 68
  personal servant,
    hiring of 63
  sending anemone roots
    to Harry Veitch 68
  travelling in rural China
    63–4
  trip to Huailai with John
    Abbott Anderson 63–5, 68
  writing to Sargent about
    trees in Peking/Huailai 68
year 1909 (expedition to
Jehol area)
  details of journey 68
  highwaymen 74
  ill-health 70
  ill-health in Dolonor 74
  Jehol elms in Cal-ceen-Wong
    71
  Luan river and Great Wall
    69
  Luan river, south of Jehol
    69
  mules and carts on road
    from Cal-ceen-Wong **71**
  plant-collecting work
    68, 70–1
  poor lodgings on the road
    74–5
  report of felling of
    Imperial hunting-forest 68
  second expedition in
    autumn/winter 74–5
  wild animals 74

234

INDEX

year 1909 (expedition to Wutai-shan)
  camp **70**
  map of area 74
  no tree seeds 75
  passport 71, 73
  with priests **73**
  trip to Pao-ting-fu, Shansi (now Baoding, Hebei Province) 73–4
year 1909 (taking stock)
  Morrison's advice about Shansi/Shensi provinces 76–7, 83
  photographs 76
  preparations for 1910 season 77–8
  quantities of seeds and specimens collected 75
  Sargent's criticism of results 75–6, 85–6
  Veitch's satisfaction with results 75
year 1910 (Shensi expedition)
  comparison with Wilson's 1910 expedition 78–9, 81
  highwaymen 91
  on horseback with Mafu **85**
  Moutan-shan and tree peonies 83, **84**, 84, 88, 91–2
  poppy cultivation and meeting Hosie in Xi'an 85
  Qinling irate pilgrims and Anderson's *sub rosa* activities 87–8, 90
  Qinling mountains 86–8, **87**, 90
  on the road to X'ian 83–6
  Sargent's criticism of collecting results 89, 90–1
  seeds and plants collected 88–9, 90, 91–2, 93
  sickness of team members 88, 90
  on the way to Shensi **80**
  wild animals 84
year 1911 (Kansu expedition)
  with armed escort **98**
  botanising around Choni 98–9
  botanising around Minchow 93, 95
  Choni Prince, cordial relationship with 96–8

Choni town, main street **97**
Christie's advice to focus on Choni 93–4
contract not extended 93, 94, 99–100, 103
Morrison's letters and report of anti-foreigner feelings 94–5
Mount Hua trip and fall 99, 147, 149
moving to Choni 95–6
passport and safety in Choni 96, 98
Peling range trip 102–3
Republican (Xinhai) Revolution (1911) 100–2, 103–6
wild animals 102
year 1912 (leaving China)
  ambush and back to Peking 104–7, **104**, 120
  last days in Peking 107–8
  meeting with Han An and Chinese Forestry Department idea 109–10
  memorandum on 'The afforestation question in China' 108–9, 110, 213–14
  return to Britain 108, 110, 111
year 1912 (taking stock)
  assessment of expedition 116–17
  cost of expedition 111
  reporting back to Harry Veitch 111
  salary extended to cover extra time in China 111–12
  Sargent's/Veitch's disappointment 111
  seedlings of *Aesculus* from Veitch to Sargent and Kew 112
  Veitch records of plant material received 112–14
  Wilson's and Purdom's collecting results compared 111, 113, 114–17
  Wilson's heroic status and Sargent's fundraising efforts 115–16
Chinese 1914–15 Farrer plant-collecting expedition

accounts of by Farrer
  assessment of accounts 131, 132–3, 140, 149, 158
  book 1 – *On the Eaves of the World* 133, 146, 149, 158, 196
  book 2 – *The Rainbow Bridge* 196
  books read by Betty Clifton 187
  Gardeners' Chronicle articles 128†, 140, 146, 149–51, 152
  Purdom's unacknowledged contribution to 149–50
  quotes from accounts 133, 136, 139, 142, 146–7, 149, 158, 195
before departure
  Farrer's meeting with Purdom 127–8, 133
  Purdom's acceptance of offer 128–9, 133
map of Purdom's movements (1914–15) **143**
year 1914 (January–October)
  arrival in Peking 133
  in camp **137**
  Chinese staff **140**
  dressed as Chinese muleteer in Chago i, **141**, 142
  Farrer's arrival in Peking 134
  Farrer's behaviour and Purdom-Farrer partnership 138–9, 142
  illness and stop at X'ian 135–6
  journey and botanising in Kansu mountains 134–7
  Mary Gaunt, meeting with 135
  Min-shan mountains, botanising on 140, 142
  Morrison, reconnecting with 134
  mule carts bogged down in mud **135**
  passports and security issues 134, 137, 139
  re-engaging same staff 133, 146
  re-tracing itinerary for seed collecting 140, 142

INDEX

reusing equipment
  kept in China   133–4
Sha-tan (Satanee)
  mountains, botanising in
                                139
Thundercrown (Leigu Shan
  mountain), botanising in
                          139, 146
Tibet, Chago and alpines
                             137–9
WWI as complete surprise
                                140
year 1914
  (November–February 1915)
  Farrer and Meyer-
    interpreter episode
                        145–7, 153
  with Farrer and Viceroy of
    Koko-nor (Lanchow)   148
  Farrer's financial deal
    with Bulley   152
  Farrer's letter to
    I. B. Balfour in
    praise of Purdom   152–3
  Farrer's room in Lanchow
                                147
  Gilbert's visit to Farrer
                          156, 184
  journey to Lanchow
    and stay   147, 149
  seed packing/sending and
    funding of expedition
                          150–1, 156
  trip to Ajiao
    mountains on his own   147
  trip to Choni on his own
                                156
  WWI and impact on
    funding of expedition
                             151–2
  WWI and Purdom's/Farrer's
    reasons for not enlisting
                             155–6
year 1915 (March–December)
                                161
  "bored stiff with
    Reginald Farrer"   161
  botanising in Gadjur
    mountains,   156–7, **157**,
                             159–60
  Farrer–Bulley financial
    deal in difficulty   160
  Farrer's praise of
    Purdom's skills   161
  fur coats, buying   160

homoeroticism in Farrer/
  Purdom relationship,
  no evidence for   158–9
Lanchow, back to/in   160
Morrison, reconnecting with
                                161
news from the war   161
return to Peking via
  Yangtze Gorges   160–1
Tien Tang monastery,
  welcome from   157, 159
trip to Ajiao mountains
  and Choni on his own   160
Wolvesden House
        157–8, **157**, **158**, 159, 160
Chinese Forestry Service
  see Forestry Adviser to
  Chinese Government
Chinese Imperial Maritime
  Customs Service   45, 46*
Chinese Labour Corps (CLC)
                             172–3
Chinese languages
  'brush conversation'   145
  Classical Chinese   145, 146
  local languages   146
  Mandarin   46, 62, 145, 146
  Purdom's knowledge
    of Chinese 46, 62–3, 146, 149
Ch'ing (Manchu) dynasty
                  45, 51–2, 53–4, 55,
                  56, 61, 67, 96, 100–1
Choni (now Joni or Zhuoni)
        94, 95–6, 98, 140, 156, 160
  Acho-che-ro (Choni resident)
                           98, **99**
  Choni town, main street   **97**
  Yang-Chi-Ching, Prince of
    Choni   94, 95, **95**, 96–8, 200
  see also Changtingssu
    monastery (nr Choni)
Chou Hsüeh-Hsi (Zhou Xuexi)
                                163
Chow Tzu-Chi (Zhou Ziqi).
                          163, 167
Christ, Hermann, Dr   90
Christian missionaries
      55, 95, 96–7, 101–2, 138, 173
China Inland Missionary
  Society   47
Christie, William
      93–4, 96–7, 102, 140, 149, 157
Chun, Prince (Regent)
                      57, 103*, 104
Ci-An (Tsu-An)   53–4

Cinchona trees   7
Cixi (Tsu-Hsi), Empress
  Dowager   53–4, 56, 57, 58, 58*
clematis   102, 159
Clifton, Baron *see* Darnley,
  Edward Bligh, 7th Earl of
  Darnley and Baron Clifton
Clifton, Elizabeth 'Betty' Bligh,
  Baroness Clifton
  aged nineteen   **186**
  aged three in coronation robes
                         185, **185**
  biographical details
    background and education
                             185–6
    character and personality
                                186
    financial situation   188
    Leveson family papers
      v, 231n162   (185)
  China
    arrival in Shanghai
                          185, 186
    birthday fancy-dress ball
                             186–7
    The Moon Flower at British
      Legation theatre   187
    stay with the Bartons and
      travelling   188–9
  Purdom
    age and social status
      differences   187–8
    knowledge of Purdom's life
                                187
    knowledge of trees/China
      (from Purdom?)   188–9
    marriage/idyll hypothesis
                             189–90
    mystery of Diamond
      Mountain trip   189, 190
    obituary of Purdom   191
    poem in Purdom's
      obituaries   190–1
  return to England
    Bar examinations with
      first-class honours 186, 197
    Daily Mail job   197
    illness   197
    Lenox Fane pseudonym
                                198
    never married   198
    sudden death   198
    writing and circle of friends
                                198

236

INDEX

writings
  articles, books and plays 188
  Legation Street (novel) 231n164 (186)
  Living in the Country 189
Cobham Hall, Kent 185, 188
Confucius 51, 103*
conifers 68, 74, 75, 78, 86, 89, 199
Conrad, Joseph 198
Conservative Party
  Balfour's resignation 33
  General Election (1906) 34
Coombe Wood nursery
  creation 14
  dove trees (*Davidia involucrata*) marketed as rarities 15†
  lease expiry date 100, 127
  plant auctions 127, 151
  'Purdom Collections' ledger 112–13, 183
  Purdom's introductions 120
  Purdom's job at 13, 16, 17
  Purdom's tree peonies 92
  red-barked birch (*Betula albosinensis*) seedlings 126
  Wellingtonia gigantea seedlings, profits from 14*
  Wilson's introductions 46, 78
  Wilson's plant material unfairly prioritised by 123
  Wilson's surplus seeds dumped by 114
  see also James Veitch & Sons
Cooper, Roland 168*
Corn Laws, repeal of (1846) 29
cotoneasters 79
Cotter, Jack 30
Country Life, Heriz-Smith's article on Purdom i
Cox, Euan 132
Crataegus 89
Craven Nursery, Ingleborough 127, 131†, 150, 152, 152*, 156
Crittenden, Fred 113
Crooks, William 'Will' 30–1, 35, 36, 41
Cultural Revolution (China, 1966–76) 202

Daily Mail, Betty Clifton working for 197

Dallimore, William
  Bedgebury Pinetum, creator of 21
  for interventionist reforestation policy 22
  Kew's Arboretum propagation pits foreman 21, 26–7
  leading authority on trees 21
  on Purdom for Sargent's China plant-collecting expedition 44
  on Purdom's dismissal from Kew 42
  on Purdom's rejection of Arnold Arboretum position 41
  on staff pay meeting with Thiselton-Dyer at Kew 26
  on Thiselton-Dyer 227n20 (23)
  unpublished memoir iv
Daphne 71
Daphne rosmarifolia 136
Darnley, Edward Bligh, 7th Earl of Darnley and Baron Clifton 185, 187*, 188
Darnley, Ivo Bligh, 8th Earl of Darnley 188
Darnley estate 188
Darwin, Charles 23
Davidia involucrata (dove tree) 15†, 45
Dawson, Jackson 39†
delphiniums 159
'Derby scheme' (1915) 155
dessicationist theory 20–1, 68‡
Deutzia 71
Devanha (steamer) 185, 186
Diamond Mountain (North Korea) 189, 190
Dipelta floribunda (formerly *Dipelta elegans*) 136, 142
Dolonor (now Dolon Nor) 74
dove tree (*Davidia involucrata*) 15†, 45
Duncan, Joseph 38
Dungan (Muslim) Revolt (China, 1862–77) 53, 54

Easter Rising (Dublin, 1916) 168
economic botany 20, 22

education (UK)
  national debate (1860s) 3–4, 5
  and Smiles' *Self-Help* book 4–5
  see also self-improvement
Education Acts (England and Wales, 1870 and 1880) 4
'Eight Power Alliance' 56
Ejder, Erland 88†
Elliott, Thomas, Sir 35, 36
elms 86
  Jehol elms 71
  Ulmus pumila 'pendula' (Peking elm) 112§, 193
Elwes, Henry John, *Trees of Britain and Ireland* (Augustine Henry and Henry John Elwes) 46
Empress of Japan (ship) 49
English cemetery, Peking 193, 202
Ensor, Robert 227n29 (29)
Eucalyptus 7
  Eucalyptus grandis 20

Fagus sylvatica 'Brathay purple' 10
Fairchild, David 122–3
Fane, Lenox (Betty Clifton's pseudonym) 198
  see also Clifton, Elizabeth 'Betty' Bligh, Baroness Clifton
Farrer, James Anson 131
Farrer, Reginald
  biographical details
    background 131–2
    Buddhism 131, 155
    character traits 132
    death in Burma 184
    health problems 131
    homosexuality 132, 158–9
    Morrison, row with 164–7
    mother's destruction of his papers 196
    will (drawn in China) 134
    WWI propaganda work 172†
  horticulture and plant-hunting
    alpine plants expert 131
    Chinese expedition *see* Chinese 1914–15 Farrer plant-collecting expedition

INDEX

Craven Nursery 127, 131†, 150, 152, 152*, 156
Dolomites expedition (1913) 128, 133
Farreria (species of *Wikstroemia*) 136
Gentiana farreri 159
Isopyrum farreri 159
'Lonely Poppy' (*Meconopsis psilonoma*) 140, **208**
orchid collector 11
'Primula hybrids in nature' (RHS Primula Conference, 1913) 127
Primula purdomii, admiration for 127, 133
Royal Botanic Garden Edinburgh collection i
Viburnum fragrans (later *V. farreri*) 136
on "Wilsonian weeds" 114*
Purdom
  correspondence with after their expedition 169–71, 174, 184
  delivery of his seeds to Harry Veitch 183
  first heard about from Sargent 127, 133
  homoeroticism, no evidence for 158–9
  news of from Nell Purdom's China letters 172, 174
  in praise of 152–3, 161
  with Purdom and Viceroy of Koko-nor (Lanchow, 1914–15) **148**
  on Purdom's fall during Mount Hua trip 99, 147, 149
  on Purdom's qualifications for Forestry Adviser position 161
  unfairly overshadowed by Farrer ii
Farreria (species of *Wikstroemia*) 136
Feng Yu-Hsiang, General 180–1
Fenwick, Walter 152
Fenwick-Owen, George 102–3, **103**, 152, 156
ferns 90

The Flora of British India 23, 33
food
  in northern and north-western China 64
  in Tibet 64
  see also inns (in China)
foot-binding 57
forestry
  deforestation in Shansi Province **86**
  dessicationist theory 20–1, 68‡
  Industrial Revolution and deforestation/reforestation debates 22
  Purdom's photos on deforestation in China 125, 126
  Purdom's 'Suggestions for training forest officers' paper 125, 215
  Purdom's 'The afforestation question in China' memorandum 108–9, 110, 213–14
  specific tree species *see under* plants
  see also Forestry Adviser to Chinese Government; Weichang Hunting Forest
Forestry Adviser to Chinese Government
  Chinese Forestry Service creation under Chow Tzu-Chi 163
  Forestry Service officers (incl. Han An) **178**
  Han An's appointment 163–4, 167
  mandate of Service 164
  number of nurseries and trees planted 183–4
  Purdom's contract (1916–19) 163, 167
  Purdom's contract (1919–23) 176–7
  Purdom's salaries 163, 177, 184
  Sherfesee's appointment 164, 165, 167, 177
  Sherfesse-Purdom conflict 166, 167
  under-employed under Ku Chung-Hsiu 167–8

Wang Ching Ch'un's contribution 177
Farrer-Morrison episode
  Farrer on Purdom's qualifications for position 161
  Morrison supporting Purdom for position 163, 164
  Morrison's row with Farrer and Purdom 164–7
  Purdom distancing himself from Farrer 167
  Purdom's loss of Morrison's support 166–7
Purdom's time as Forest Adviser
  Chinese awards in recognition of his work 184
  forestry stations 179–81, **180**
  Mount Jigong forest station (now Purdom Forest Park) 180–1, **209**, **210**
  nurseries/forestry training schools scheme with Han An 177–80
  Peking-Hankow (Kin-Han) line scheme with Han Ann 177, 179
  private rail-car as home and office 179, **179**
  prose poem about plant-hunting in China 171, 216
  reforestation in Chihli (Zhili) province with Han An 168
  relationship with Joseph Hers 181–2, 184
  sending plant material to Arnold Arboretum 182–3, 184
  sending seeds to Harry Veitch 183
  sending seeds to Kew 183, 184
  sketch of study in Peking (1917) **170**
  status in top layer of expat society 183
  tea party in Peking (1917) **169**
  tours of inspection 174–5
  trips for tree nursery projects 164, 167, 175
  trips with Han An 175

INDEX

under-employed and depressed 168–71, 172, 175†
visit from sister Nell 171–2, **173**, 174, 175‡, 176*, **176**
work for British secret services (possibly) 175–6
writing to Farrer again (sometimes) 169–71, 174, 184
WWI news and not joining up 168–9, 170–1, 172–4, 183
Forrest, George ii, 43–5, 47, 138, 171, 175*
France
  Annam kingdoms (now Vietnam), control over 55
  Canton, battle of (1857–58) 53
  Old Summer Palace (Peking), burning of 53
  treaties with China (19th century) 54
Free Grammar School (Ambleside) 4

Galsworthy, Jane 202
Gansu *see* Kansu (now Gansu) province
The Garden (RHS magazine), letter re. Will Purdom 219
gardeners
  wages 25, 26, 29–30
  see also British Gardeners' Association; Royal Botanic Gardens, Kew
The Gardeners' Chronicle
  Coombe Wood plant auctions, news of 127
  Farrer's articles 128†, 140, 146, 149–51, 152
  gardening profession, overcrowding in (letters) 29–30
  Purdom's article on *Aesculus Chinensis* 121, 122
  Purdom's articles on plant-collecting in China 126, 168
  Purdom's father's subscription to 4
  Purdom's obituary 196
  Purdom's photograph dressed as Chinese muleteer 142†

Royal International Horticultural Exhibition (1912), reports of 119
Thiselton-Dyer's letter re. gardeners' wages and response from 'Old Kewite' 29
Garnett, Jack 88
Gathorne-Hardy, Alfred 161
Gaunt, Mary, on meeting Purdom in China (*A Broken Journey*) 135
General Election (UK, 1906) 33–4
gentians 159
Gentiana farreri 159
Gibbs, Vicary 119
Gilbert, Ronald 149, 156, 184
Giles, John 58‡
Ginkgo biloba 181‡
Government Workers Federation (GWF) 27, 28, 31, 34
grain prices (1870s–80s) 29
Grand Hotel des Wagons-Lits (Peking) 62, 67†, 161
Great Wall of China 68, **69**, **80**, 84, 109
Groombridge, Amos 47
Groombridge, Annie Valentine 46–7, 120–1, 123, 198
groundsel 99
Gu Zhongxiu *see* Ku Chung-Hsiu (Gu Zhongxiu)
Guangxu Emperor *see* Kwang-Hsu (Guangxu) Emperor

Han An (Han Ngen)
  correspondence with Purdom 125†, 125–6, 128
  first meeting with Purdom 109–10
  Forestry Service work with Purdom 163–4, 167, 175, 176, 177, 179, 180–1
  with other Chinese Forestry Service officers **178**
  post-1922 positions and death 201
  in Purdom Forest Park visitor centre paintings 201
Han Suyin (pen-name of Rosalie Chou) 181†, 228n56 (67)
Hardie, Keir 34

harebell poppies 160, **205**
Harrow, George 17
Hart, Robert, Sir 45, 46, 58
hellebores 136
Helleborus thibetanus 136
Hengduan mountains (western Szechuan), plant biodiversity 79, 81
Henry, Augustine, Dr 45–6, 58, 59, 189n‡
Trees of Britain and Ireland (Augustine Henry and Henry John Elwes) 46
Henty, George i
herbaceous plants *see under* plants
herbarium specimens, preparation process 49
Herbert, Aubrey 11
Heriz-Smith, Shirley i
Hers, Joseph 175, 181–2, 184, 199
  photograph of **182**
  photograph of *Aesculus chinensis* by **106**
  plants
    Acer hersii 181
    collection of over 6,000 herbarium specimens 199
    Syringa pubescens julianae 'Hers' 182
Hevea (rubber tree) 7, 12
highwaymen 64, 74, 91, 98, 103, 104, 105–6
homosexuality 132, 158–9
honeysuckle 98
Lonicera caerulea (honeyberry) 98*, 126
Hooker (family) 36
Hooker, Joseph, Sir 6, 7, 12, 22, 36
Sir Joseph Hooker prize (Kew) 21
Hosie, Alexander, Sir
  etymology of *Aesculus Chinensis* tree's Chinese name 121
  herbarium specimens to Kew 59
  Purdom meeting with in Xi'an 85
  Purdom travelling with to Peking 61, 62, 66
Purdom–Han An first meeting 109

239

INDEX

Purdom's letter of
introduction to   47, 59
Soothills, relationship with
   107, **107**, 108
Sowerby's 'Shensi Relief
Column' episode   101
Hough, Charles, Dr
   128, 149–50, 151, 197
White Craggs
   124, **124**, 127, 197
Hsi-Sheng Ch'i   176
Huailai, Purdom's trip to
with Anderson   63–5, 68
Hugh Low & Co nursery
(Bush Hill, Enfield)
   10, 12*, 12–13, 17, 25
hypericums   111

Imperial Defence Staff   64, 65
Imperial hunting-grounds
see Weichang Hunting Forest
Imperial Maritime Customs
Service see Chinese Imperial
Maritime Customs Service
Index Kewensis   23
India
   Cinchona tree plantations
   for quinine   7
   Eucalyptus grandis,
   propagated by Kew   20
   The Flora of British India
      23, 33
   Kew's links with
   botanical gardens in   6
   Royal Botanic Gardens,
   Calcutta   20, 33
   see also British Army in India
India Office   7
Industrial Revolution   19, 22, 52
Ingleborough Hall/Estate
   131–2, 139
   Craven Nursery
   127, 131†, 150, 152, 152*, 156
   harebell poppy (1950s)   **205**
inns (in China)
   63–4, 74, **80**, 136, 189
iris   159
Irving, George   198
Isopyrum farreri   159

James Veitch & Sons
   creation and management
   issues   14–16
   exhibits

Chelsea Flower Show
   Aster flaccidus var.
   purdomii exhibit   125
   RHS Primula Conference
   Primula purdomii exhibit
      127
   Royal International
   Horticultural Exhibition
   stand   119–20
nurseries
   Coombe Wood see Coombe
   Wood nursery
   Langley   14, 92, 113
   Royal Exotic Nursery
   (Chelsea)
      13–14, 75, 91, 92
seeds and plants
   from Augustine Henry   45
   from Chinese 1909–12
   expedition   48, 75
   New Hardy Plants from
   Western China (introduced
   through Mr. Ernest Wilson)
      119
   'Seeds collected by
   W Purdom' register   112
   Wellingtonia gigantea,
   introduction of to Britain 9
   see also Veitch, Harry
Japan
   Korea, control over   55
   Shadong 'concession' given to
      180
   treaties with China
   (19th century)   54
Java, Cinchona tree plantations
for quinine   7
Jehol (modern Chengde)
   47, 62, 68, 71, 74, 89, 109
Jehol elms   71
'Jesuit's bark' (quinine)   7
Johns, William   25, 156
Joni see Choni (now Joni or
Zhuoni)
Jordan, John, Sir
   61, 62, 63, 101, 106†, 199
Journal of the Arnold
Arboretum, notice of
Purdom's death   196
Journal of the Kew Guild
   Purdom's obituary   191
   Wilson's contributions to   59*
Jung-Lu (Rong-Lu)   54
junipers   99

Kansu (now Gansu) province
   language   146
   northern Kansu
      deforestation   149
      wildlife   102
Purdom's 1911 expedition
see under Chinese 1909–12
Sargent–Veitch plant-
collecting expedition
Purdom's 1914 expedition
see Chinese 1914–15 Farrer
plant-collecting expedition
Kansu mole (Scapanulus oweni)
   102
Kenya, translocation of
Eucalytus species from Australia
   7
Kew botanic gardens
see Royal Botanic Gardens, Kew
Kew Employees Union
   27–8, 30–1, 34, 36,
   37, 39, 41, 46
Kew Guild
   7–8, 28–9, 38, 81, 120
Journal of the Kew Guild
   59*, 191
Keyte, John Charles   101
King, James   34
Kingdon-Ward, Frank
   65*, 93–4
Kin-Han (Peking-Hankow)
railway   172, 173, 174–5,
   175*, 177, 179, 202
Kipling, Rudyard   i
Kitchener, Herbert, Earl   168–9
Kodak 'pocket' camera   46, 73†
Koko-nor   137
   Koko-nor lake   149
   Zhang Bing Hua, Viceroy of
   Koko-nor   **148**, 149, 157, 160
Korea
   Diamond Mountain
   (North Korea)   189, 190
   Japan's control over   55
Kozlov, Pyotr   93†
Ku Chung-Hsiu
(Gu Zhongxiu)   168
Kuomintang   126, 200
Kwang-Hsu (Guangxu)
Emperor   54, 57, 62

Labour Party
   General Election (1906)   34
   and Kew trade union
   activities   27, 30–1, 36, 39
   Purdom's association with 41

240

INDEX

Labrang monastery 99
Lake District 9, 120
Lake District Herald, Purdom's obituary 195–6
Lanchow, Farrer's and Purdom's stay in 147, **147**, 149, 160
Lanchow-Lianyungang (formerly Lanzhou-Lunghai) rail line 175, 181
Langley nursery 14, 92, 113
larches 43, 75, 77, 90, 91
Leigu Shan mountain (Thundercrown) 139, 146
Leuw, Johannes de 145
Leveson (family) v
Leveson, Arthur, Rear-Admiral (Betty Clifton's stepfather) 185, 188
Leveson, 'Dolly' McLeod (Betty Clifton's aunt) 185, 186
Leveson, Jemima, Lady (Betty Clifton's mother) 185, 188, 189, 197–8
Li Hung-Chang (Li Hongzang) 56
Li Yuanhong 100
Liberal Party, General Election (1906) 33, 34
lilac *see Syringa* (lilac)
Lilium regale (Easter lily) 78–9, 89
Lin Ze-Xu 52
Lobb, Thomas 13
Lobb, William 9‡, 13, 14*
loess 135*
London County Council 28, 30, 34, 38
London plane (*Platanus x Acerifolia*) 181
'Lonely Poppy' (*Meconopsis psilonomma*) 140, **208**
Lonicera see honeysuckle
Low, Hugh, Sir 12–13
see also Hugh Low & Co nursery (Bush Hill, Enfield)
Low, Stuart 12
Luan river
Great Wall from **69**
south of Jehol **69**
Lunghai rail line *see* Lanchow-Lianyungang (formerly Lanzhou-Lunghai) rail line

Ma Ch'i (Ma Qi), General 96
Ma Fu-hsiang (Ma Fuxiang), General 200
MacGregor, Donald 58, 61
magic-lantern slides, Purdom's lectures with 125
malaria 7
Malay peninsula, rubber plantation 12
Malus tree 84
Manchu (Ch'ing) dynasty 45, 51–2, 53– 4, 55, 56, 61, 67, 96, 100–1
Manchurian water rice (*Zizania latifolia*) 111
Mandarin 46, 62, 145, 146
Mao Tse Tsung 202
maps
of China 63, 65, 117
of Purdom's movements (1909–12) **72**
of Purdom's movements (1914–15) **143**
Matthew, John 9‡
May the Fourth Movement (China, 1919) 180
Meconopsis (poppies) 93, 94, 95, 126
Meconopsis psilonomma ('Lonely Poppy', now *Papaver psilonomma*) 140, **208**
Meconopsis quintuplinervia 139
Merriman, George 31
Meyer, Frank
death 175
dispute with Chinese interpreter (1914 expedition) 145–7, 153
photograph **121**
Purdom
letter to Fairchild re. Purdom's time in China 122–3
visit to in Westmorland 121–3
Sargent
dealings with 43, 48, 75
disappointment in Purdom's results mentioned by 121, 122, 123
Theodore Roosevelt, meeting with 21

University of California Davis Library collection iv
Wilson
briefed by 59
unfairly overshadowed by 116
Military Service Act (UK, 1916) 171
Mill, John Stuart 19
Minchow (now Minxian), Kansu (now Gansu) province 93–4, 95, 102–5, 107, 111, 116, 136, 140
Ming dynasty 51, 53
Min-shan mountains 137, **137**, 139, 140, 142
missionaries *see* Christian missionaries
Mitchell Library (State Library of New South Wales), Dr George Ernest Morrison Collection iv, 230n140 (166)
monasteries
Chago 137, 138, 139, 142
Changtingssu (nr Choni) 95, 102
Labrang 99
Tien Tang 157, 159
monkey-puzzle tree (*Auracaria*) 13
moraine 9
Morrison, George Ernest
death 199
diary-keeping method 230n140 (166)
Farrer, row with 164–7
marriage 187*
Mitchell Library Morrison collection iv, 230n140 (166)
and Purdom
advice to Purdom about Shansi/Shensi provinces 76–7, 83
on Purdom being "bored stiff with Reginald Farrer" 161
on Purdom being grossly overpaid 199
Purdom on anti-foreigner sentiment in China 95
Purdom on deforestation of Weichang Hunting Forest 77, 94
Purdom on his Sargent–Veitch contract not being renewed 103

INDEX

Purdom reconnecting with (1914) 134
Purdom reconnecting with (1915) 161
Purdom's loss of Morrison's support 167
Purdom's 'Suggestions for training forest officers' paper 125–6, 215
Purdom's 'The afforestation question in China' memorandum 108–9, 110, 213–14
support for Purdom as Director of Chinese Forestry Service 163, 164
telegram to London about Purdom's ambush 106†
on Sherfesee 164
Special Adviser to Chinese government 125, 198–9
The Times correspondent in Peking 58, 76, 100†, 125
on Xi'an massacre (1911) 100†
Mount Hua (Qinling range), Purdom's fall 99, 147, 149
Mount Jigong forest station *see* Purdom Forest Park, Xinyang (formerly Mount Jigong forest station)
Mount Tabai (Qinling range) *see* Tai-Pei-Shan (now Taibai Shan, Qinling range)
Moutan peonies *see* tree peonies (Moutan peonies)
Moutan-shan (Shensi (now Shaanxi) province) 48*, 83, 84, **84**, 88
Muslim (Dungan) Rebellion (China, 1862–77) 53, 54
Mutual Improvement Movement 5
Mutual Improvement Society (for Kew student gardeners) 7, 21

Nanking, Treaty of (1842) 52, 54
Natural History Museum, London
  designed by Alfred Waterhouse 3*
  Duke of Bedford expedition 93, 102
  Fenwick-Owen expedition 102
Naval, Shipping and Fisheries exhibition (Earls Court, 1905) 30
New Hardy Plants from Western China (introduced through Mr. Ernest Wilson) 119
Ng Achoy *see* Wu Ting-Fan, Dr (Ng Choy or Ng Achoy)
Ng Choy *see* Wu Ting-Fan, Dr (Ng Choy or Ng Achoy)
Nian Revolt (China, 1851–68) 53, 54
North China Herald, Purdom's obituary 195

Oceanic (ship) 47
Old Summer Palace, Peking
  burning of (1860) 53
  suitability for tree nursery site 167
Onslow, William, 4th Earl of Onslow 27, 28, 33
opium trade/wars 52–3, 57, 85, 103
orchids
  orchidomania 10–11
  rivalry and high mortality amongst orchid-hunters 11–12
  supplied by Hugh/Stuart Low & Co nursery 10, 12
  from Thomas Lobb's expedition for Veitch 13
Ordos desert (Shensi province) 84, 109

Paine, William 27
Pao Ma Chang racecourse (now Lianchuachi Park, Peking) 67
Pao-ting-fu, Shansi (now Baoding, Hebei Province) 73–4
Paulownia tomentosa var. tsinglingsensis (Pai) Gong Tong (formerly *Paulownia glabrata*) 85*, 86
pear trees (*Pyrus*) 84, 86, 88, 89, 93, 113
Pearson, Constable 133, 134
Peking
  description of city 61–2
  English cemetery 193, 202
  Forbidden City 51*, 61, 62, 67
  foreign residents 67–8
  Legation Quarter 56, 62, 67†, 187, 189
  looting of by 'Eight Power Alliance' (1900) 56
  motor race to Paris (1907) 61*
  Old Summer Palace 53, 167
  Pao Ma Chang racecourse (now Lianchuachi Park) 67
  population figures (ca. 1909) 61
  Temple of Heaven **107**, 108, 167
  Tsin Shih Tse temple **106**
Peking Daily News, Purdom's obituary 190
Peking elm (*Ulmus pumila* 'pendula') 112§, 193
Peking Times, Nell Purdom's notice of thanks for floral tributes 193
Peking-Hankow (Kin-Han) railway 172, 173, 174–5, 175*, 177, 179, 202
Peling mountain range 48, 102–3, 126
  Tibetan peasant girl **76**
peonies
  herbaceous 70, 102, 139
  tree peonies (Moutan peonies) 48, 83, 84, 88, 91–2, 93, 126, 136
    Paeonia rockii 92†
    Paeonia suffruticosa 92
    Paeonia veitchii var. *woodwardia* 92
Peru, seeds of *Cinchona* trees from 7
Philadelphus pekinensis 182–3
photographs (taken by Purdom) 76
  Tibetan peasant girl **76**
  see also cameras
pines 48, 68, 75, 91, 99, 183
  lacebark pines 84
  Pinus koraiensis 189n‡
Plagiospermum sinense 89
Plantae Wilsonianae 85*, 92, 112
plants
  herbaceous plants
    aconites 111
    Alhagi maurorum

242

INDEX

(camel-thorn) 20
alliums 183
alpines 127, 128, 131, 137, 159
androsaces (rock jasmine) 150
anemones 68, 70–1, 113, 139
Aster flaccidus var. *purdomii* 125
delphiniums 159
ferns 90
Gentiana farreri 159
gentians 159
groundsel 99
hellebores 136
iris 159
Isopyrum farreri 159
Lilium regale (Easter lily) 78–9, 89
Meconopsis (poppies) 93, 94, 95, 126
Meconopsis psilonomma ('Lonely Poppy', now *Papaver psilonomma*) 140, **208**
Meconopsis quintuplinervia 139
orchids 10–12, 13
peonies 70, 102, 139
polygonum 111
poppies 139, 150, 159
poppies, harebell poppies 160, **205**
poppies, *Papaver psilonomma* ('Lonely Poppy', formerly *Meconopsis psilonomma*) 140, **208**
potentilla 95
primroses 139, 150
Primula maximowiczii 95
Primula purdomii 127, 133
primulas 74, 79, 93, 94, 98, 111, 126, 139, 159
Rodgersia aesculifolia 'Purdomii' (formerly *Rodgersia purdomia*) 202
rose family 79
saxifrage 159
Senecio 126
Zizania latifolia (Manchurian water rice) 111

shrubs
abelias 84, 111
berberis 139
Buddleia alternifolia 136, 142, 150–1
Buddleia purdomi 136
Caragana (Siberian pea tree) 71
clematis 102, 159
cotoneasters 79
Crataegus 89
Daphne 71
Deutzia 71
Dipelta floribunda (formerly *Dipelta elegans*) 136, 142
Farreria (species of *Wikstroemia*) 136
honeysuckle 98
honeysuckle, *Lonicera caerulea* (honeyberry) 98*, 126
hypericums 111
Paeonia (Moutan/tree peonies) 48, 83, 84, 88, 91–2, 93, 126, 136
Paeonia rockii 92†, 136‡
Paeonia suffruticosa 92
Paeonia veitchii var. *woodwardia* 92
Philadelphus pekinensis 182–3
Plagiospermum sinense 89
Rhododendron purdomii 88, 196, 198, **206**, **207**
rhododendrons 12, 13, 68, 74, 79, 84, 139
Rosa Moyessi 120*
Rosa Willmottiae 120*
rose family 79
roses 183
sea buckthorns 84
Viburnum fragrans (later *V. farreri*) 136
viburnums 84
trees
Acer hersii 181
Acer tegmentosum 181§
Aesculus chinensis 89, **106**, 107–8, 111, 112, 121, 122, 202, **212**
Auracaria (monkey-puzzle) 13

Betula albosinensis (red-barked birch) 91, 126
birches 10, 48, 68, 183
chestnuts 68
Cinchona 7
conifers 68, 74, 75, 78, 86, 89, 199
Davidia involucrata (dove tree) 15†, 45
elms 86
elms (Jehol elms) 71
elms (Peking elms) 112§, 193
Eucalyptus 7
Eucalyptus grandis 20
Fagus sylvatica 'Atropurpurea' 10
Ginkgo biloba 181‡
Hevea (rubber tree) 7, 12
Jehol elms 71
junipers 99
larches 43, 75, 77, 90, 91
Malus 84
Paulownia glabrata 85*, 86
pines 48, 68, 75, 84, 91, 99, 183
pines, lacebark pines 84
Pinus koraiensis 189n‡
Platanus x Acerifolia (London plane) 181
poplars 68, 84, 86, 199
Populus cathayana 182–3
Prunus 68, 113
Prunus humilis 183
Prunus Padus var. *pubescens* forma *Purdomii* 70*
Prunus triloba 197
Prunus triloba "Multiplex" 183
Pyrus (pear trees) 84, 86, 88, 89, 93, 113
rose family 79
sorbus 79
sumac trees 183
Syringa (lilac) 84
Syringa pubescens julianae 'Hers' 182
Taxodium distichum (American swamp cypress) 180–1, **209**
Ulmus pumila 'pendula' (Peking elm) 112§, 193

INDEX

Wellingtonia gigantea   9, 14*
willows   68, 199
Platanus x Acerifolia
(London plane)   181
Plum-Tree Hall (Heversham, Westmorland)   1
poems
    Betty Clifton's poem about Purdom   190–1
    Purdom's prose poem about plant-hunting in China   171, 216
Polo, Marco   61
polygonum   111
the poor, 'deserving' vs 'undeserving'   5
poplars   68, 84, 86, 199
    Populus cathayana   182–3
poppies   139, 150, 159
    harebell poppies   160, **205**
    Papaver psilonomma ('Lonely Poppy', formerly *Meconopsis psilonomma*)   140, **208**
    see also Meconopsis (poppies)
poppy cultivation   85
    see also opium trade/wars
Populus cathayana   182–3
potentilla 95
Prain, David, Colonel
    appointment as Kew Director   33
    Kew staff dispute   34–7
    portrait   **33**
    Purdom, character reference for   44, 47
    Purdom, dismissal of   41–2
    Purdom, letter of introduction to Hosie for   59
    Purdom, recommendation of for Arnold Arboretum post   39–40, 41
    Purdom's unhappy memory of   120
    son killed in WWI   161
Presant, Victoria   219
primroses   139, 150
Primula RHS Conference (London, 1913)   127, 133
primulas   74, 79, 93, 94, 98, 111, 126, 139, 159
    P. maximowiczii   95
    Primula purdomii   127, 133

Prunus trees   68, 113
Prunus humilis   183
Prunus Padus var. *pubescens* forma *Purdomii*   70*
Prunus triloba   197
Prunus triloba "*multiplex*"   183
Przewalski, Nicholas   93
Pu Yi, Emperor of China   57, 104
Public Record Office   i
Purdom, Alan (Purdom's great-nephew)   iv, 219
Purdom, Annie (Purdom's sister)   **2**
Purdom, Gilbert (Purdom's brother)   129, 155
Purdom, Harry (Purdom's brother)   **2**, 3, 155
Purdom, Jane (Purdom's mother)   1, **2**, 3, 46, 169, 198
Purdom, Margaret 'Peg' (Purdom's sister)
    Annie Groombridge, friend of   47
    attending well-known boarding school   3
    family portrait at Brathay Lodge garden (1886)   **2**
    Meyer's visit to Purdom, concerns over   122, 123
    Purdom's letters about returning to Britain (1912)   104–5
    Purdom's London bank account managed by   49†, 171
    Purdom's stay with (1913)   125
    secondary-school teaching career in London   47, 171
Purdom, Nellie 'Nell' (Purdom's sister)
    biographical details
       Annie Groombridge, friend of   47
       marriage and work in Malaya   198
       retirement and death in Australia   198
       teaching job in Singapore   198
    China
       on Purdom hiring armed escort   105
       with Purdom in Peking   171–2, 174, 175‡, 176*

       on Purdom meeting a "Mr Christiansen"   157*
       on Purdom meeting Mrs and Dorothea Soothill   107*
    Purdom's death   190, 193, 197
    with red setter 'Mustard' (1918)   **173**
    with red setter 'Mustard' and Purdom (1918)   **176**
    Sweet Stone Bridge (life with Purdom, 1917–18) 198
    teaching jobs   179, 198
Purdom, Will
    beginnings
       birth in Plum-Tree Hall gardener's cottage   1
       childhood at Brathay Lodge (gardener's cottage)   **2**, 3
       Elementary Certificate in botany   4
       family background   1, 3
       Free Grammar School (Ambleside)   4
       training at Brathay Hall   8, 9, 10
    biographical details
       "ambush" incident (March 1912) 106, 107, 120
       Annie Groombridge and breaking off of engagement   46–7, 120–1, 123, 198
       Betty Clifton *see* Clifton, Elizabeth 'Betty' Bligh, Baroness Clifton
       British secret services, possible links with   88, 117, 175–6, 197*
       Chinese name (Pao Er Deng)   63
       Chinese speaker   46, 62–3, 146, 149
       fall on Mount Hua   99, 147, 149
       London bank account   49†, 171
       not enlisting in WWI   155–6, 171, 172–4, 183
       not entitled to vote   156
       sources for biography   i–ii, iv–v, 219
    career
       Hugh Low & Co nursery   10, 12–13, 25

INDEX

James Veitch & Sons' Coombe Wood nursery 13, 16, 17
Kew *see under* Royal Botanic Gardens, Kew
rejection of Arnold Arboretum position 39–41
work in China *see* Chinese 1909–12 Sargent–Veitch plant-collecting expedition; Chinese 1914–15 Farrer plant-collecting expedition; Forestry Adviser to Chinese Government
career break (in London after 1909–12 expedition)
efforts not recognised by Harry Veitch 119–20
failure to attend Kew Guild annual dinner 120
fortnight in Ambleside with family and friends 120
last dealings with Harry Veitch 120
move from London to parents' home in Westmorland 120
state of mind after ambush in China 120
career break (in Westmorland after 1909–12 expedition)
breaking off of engagement to Annie Groombridge 120–1, 123
Chelsea Flower Show and *Aster flaccidus var. purdomii* 125
Coombe Wood plant auctions without mentions of his introductions 127
depressed state of mind 123, 125
Farrer, meeting with 127–8, 133
Farrer's offer of joining China expedition, accepted 128–9, 133
Gardeners' Chronicle article on *Aesculus Chinensis* 121, 122
Gardeners' Chronicle articles on plant-collecting in China 126, 168
lectures to local groups with magic-lantern slides 125

Meyer's visit 121–3
Morrison and 'Suggestions for training forest officers' paper 125, 215
recognising unfair treatment from Sargent/Veitch 123
work on White Craggs 124
working overseas and social mobility 129
character and looks
clothes, taste for 66, 160, 184
cocktail-party chatter, dislike of 171
commitment to task in hand 116
cool head 116
depression 123, 125, 168–71, 172, 175†
draughtsmanship, no talent for 65†
Farrer on 152–3, 161
handsome 159
height 17
homosexuality, no evidence of 159
hunting 102*
letter-writing, poor at 117, 169
modesty, not inclined to 41
Prain's character reference 44, 47
reticence 74–5, 117, 168
self-portraits, weakness for 66*
smoker 16, 159
taciturnity 106, 117
Watson on 47
death
fateful surgery 193, 197
funeral in Peking's English cemetery 193
grave under Peking elms 193, 202
memorial (British Legation chapel/British Ambassador's garden) 191, 202, **211**
memorial stele (Purdom Forest Park, Xinyang) **210**
memorial stele (Purdom Forest Park, Xinyang), transcript and translation 189§, 190, 201, 217–18

mourned by daughter of an earl 3, 190–1
Nell's notice of thanks in *Peking Times* 193
notice of death in *Journal of the Arnold Arboretum* 196
obituary in Ambleside parish magazine 195
obituary in *Journal of the Kew Guild* 191
obituary in *Lake District Herald* 195–6
obituary in *North China Herald* 195
obituary in *Peking Daily News* 190
obituary in *The Gardeners' Chronicle* 196
as sad loss to British horticulture 3, 8, 196
photographs
1886 family portrait at Brathay Lodge garden (aged 6) 2
1901 aged 21 (studio portrait) **15**, 16
1909 self-portrait (Peking) **66**
1909 with Wutai-shan priests **73**
1910 at Moutan-shan camp **84**
1910 on horseback with Mafu **85**
1910 with Chinese innkeeper/staff **80**
1911 with armed escort (Kansu) **98**
1912 with Soothills and Hosie (Temple of Heaven, Peking) **107**
1914 dressed as Chinese muleteer i, **141**
1914 in camp on Min-shan range **137**
1914–15 with Farrer and Viceroy of Koko-nor **148**
1917 tea party in Peking **169**
1918 with Nell and red setter 'Mustard' (Peking) **176**
1919 by Great Wall gate **80**

INDEX

plants
  Aster flaccidus var. purdomii   125
  Buddleia purdomi   136
  plants introduced and collected for Arnold Arboretum   196
  plants named *purdomii*   196
  plants named *purdomii* by Royal Botanic Garden Edinburgh   170
  Primula purdomii   127, 133
  Prunus Padus var. *pubescens* forma *Purdomii*   70*
  Rhododendron purdomii   88, 196, 198, **206**, **207**
  rhododendrons, affinity for   12
  Rodgersia acerifolia var. purdomii (formerly *Rodgersia purdomia*)   202
  trees, love for/specialist knowledge of   10, 21–2, 40
  Wellingtonia gigantea, introduction to China   9
  woody plants, talented propagator of   12, 21, 40
views
  China and Chinese, affection for   125, 129
  contributing to botany vs enlisting for WWI   155
  dessicationist theory, belief in   21, 68‡
  "England seems flat"   125
  interventionist reforestation policy, support for   22
  Labour Party, association with   41
  opium in Kansu, concerns over   103
  racist attitudes, rejection of   46, 68, 116
  social justice, strong sense of   25, 46
  socialist beliefs   3, 153, 156
  Weichang Hunting Forest devastation, concerns over   77, 94
writings
  'The afforestation question in China' memorandum   108–9, 110, 213–14

Gardeners' Chronicle article on *Aesculus Chinensis*   121, 122
Gardeners' Chronicle articles on plant-collecting in China   126, 168
notebooks (RHS archives)   i
prose poem about plant-hunting in China   171, 216
'Suggestions for training forest officers' paper   125, 215
Purdom, William (Purdom's father)
  death and tombstone   198
  family portrait (Brathay Lodge garden)   2
  gardener at Plum-Tree Hall   1
  Gardeners' Chronicle subscription   4
  head gardener at Brathay Hall   3, 5, 10
    James Abbott's apprenticeship   5–6, 8, 17
    son Will's apprenticeship   8, 9, 10
  schooling and social mobility for his children   3–4, 5
  senior churchwarden at local church   3
  and son Will
    growing plants introduced by in his garden   198
    Will's death, impact on   198
    Will's obituary in scrapbook beside Betty Clifton's photo   190
    Will's visit before China expedition   46
Purdom Forest Park, Xinyang (formerly Mount Jigong forest station)   180–1, 201, **209**
  Purdom's memorial stele   **210**
  transcript and translation of   189§, 190, 201, 217–18
Purdon, Lt-Colonel Richard   173*
Pyrus (pear trees)   84, 86, 88, 89, 93, 113

Qinling mountains (Tsingling mountains, Shensi province)   79, 86–8, **87**, 116
  see also Mount Hua (Qinling range); Tai-Pei-Shan (now Taibai Shan, Qinling range)
quinine ('Jesuit's bark')   7

racism, amongst expat community   46, 67–9, 116
Redman, George   150, 152*, 156
Redmayne (family)   3–4
Redmayne, Giles   3*, 17, 124
Redmayne, Hugh   3, 10
Redmayne, Katherine Mary (Mrs)   3, 10
Regal lily (lilium regale)   78–9, 89
Rehder, Alfred   70*, 92, 181, 196
Reisner, John H.   183–4
Repton, Humphrey   185
Republican (Xinhai) Revolution (China, 1911)   100–2, 103–6, 116, 125–6
rhododendrons   12, 13, 68, 74, 79, 84, 139
  Rhododendron purdomii   88, 196, 198, **206**, **207**
RHS *see* Royal Horticultural Society (RHS), London
Ridderlöf, Stern   88†
robbers *see* highwaymen
Robin, Jennie   187*
Rock, Joseph   98, 200
  Paeonia rockii   92†
rock jasmine (androsaces)   150
Rockhill, William   62
  Land of the Lamas   62, 77–8
Rodgersia aesculifolia 'Purdomii' (formerly *Rodgersia purdomia*)   202
Rommen, Jens, Reverend   95*
Rong-Lu *see* Jung-Lu (Rong-Lu)
Roosevelt, Theodore 'Teddy'   21
rose family   79
roses   183
  R. Moyessi   120*
  R. Willmottiae   120*
Rothschilds Bank   47
Royal Botanic Garden Edinburgh
  George Forrest plant-collector for   43, 44*
  number of male staff enlisting in WWI   151*
  plants named *purdomii*   170
  Reginald Farrer collection   i
  Rodgersia acerifolia var. purdomii for Purdom's memorial   202
Royal Botanic Gardens, Calcutta   20, 33

INDEX

Royal Botanic Gardens, Kew
  about Kew
    administration and budget 19
    Aesculus chinensis from Purdom 112, 202, **212**
    Arboretum 21, 26, 30, 40, 44
    archives i
    botanical gardens across British Empire, links with 6–8, 20
    cultural values and mission 19–20
    forestry and dessicationist theory 20–1
    Herbarium 18‡, 19, 45
    Index Kewensis 23
    Kew Gardens Constabulary 12, **22**, 23
    plant identification 78, 89, 112–13
    'Seeds collected by W Purdom' register 112–13
    seeds from Purdom in China (1921) 183, 184
    staff numbers 19
  botany and plant translocation projects
    Alhagi maurorum from Cairo to Australia 20
    Cinchona from Peru to India 7
    economic botany 20, 22
    Eucalyptus grandis to Africa and India 20
    Eucalytus from Australia to Kenya and Uganda 7
    Hevea from Latin America to south-east Asia 7
  connections
    Alexander Hosie 59
    Augustine Henry 45
    Emil Bretschneider 83
    Ernest Wilson 59
    Hugh Low 12
    Joseph Hers 182
  Directors
    Hooker, Joseph, Sir 6, 7, 12, 22, 36
    Prain, David *see* Prain, David, Colonel
    Thiselton-Dyer, William 17, 22–3, **22**, 26, 27–9, 31, 33, 36, 59
  gardening training course
    application criteria and procedure 6
    application form 17–18, 36, 37, 39
    course length and structure 6, 18–19
    employment prospects 6
    employment prospects overseas 8
    evening lectures, students' dissatisfaction with 25–6
    Kew Guild 7–8, 28–9, 38
    Mutual Improvement Society 7, 21
    student gardeners' wages 25
    student gardeners' working hours 34†
  staff/students' grievances (round 1)
    foremen's wage and status demands 26
    labourers' and students' low wages 25
    petitions for higher pay 26, 28
    petitions for higher pay and Treasury's response 29–30
    setting up and involvement of Kew Employees Union 27–8, 30–1
    Thiselton-Dyer's response to employees' demands 27–9
    Thiselton-Dyer's sacking of Union representatives and resignation 31, 33
  staff/students' grievances (round 2)
    appointment of Lord Carrington at Board of Agriculture 33, 34
    appointment of Prain as Director 33, 36–7
    Kew Employees Union's memorandum to Carrington 34
    Kew gardeners' memoranda criticizing Prain's response 35–6
    Prain's response to memorandum to Carrington 34–6
    staff grievances partly addressed by HM Treasury 35, 36
    student gardeners' grievances unaddressed 36
    student gardeners' opposition to Prain's application of fixed-term rule 36–7
    student gardeners' student gardeners' 'strike' over wages 37–9, 41
  Will Purdom at Kew
    application form and letter 17–18
    Arboretum sub-foreman 21–2, 26–7, 40
    Arnold Arboretum position, rejection of 39–41
    conflicts with management and union activities 27–8, 30–1, 34, 36, 37, 41, 42, 46
    determined and intelligent, recognised as 19
    dismissal and reinstatement by Thiselton-Dyer 31, 59
    dismissal by Prain and Carrington 41–2
    making his mark 8
    Naval, Shipping and Fisheries exhibition, visit to 30
    open-ended contract argument 37, 39
    rotation through departments 20
    salary per annum 40
    sharing a room to save money 25
    Sir Joseph Hooker prize 21
  see also Kew Employees Union; Kew Guild
Royal Chelsea Hospital, London, Royal International Horticultural Exhibition (1912) 119–20
Royal Exotic Nursery (Chelsea) 13–14, 75, 91, 92
Royal Horticultural Society (RHS), London
  first Chelsea Flower Show

INDEX

(1913) 120, 125
Great Spring Shows 119
Harry Veitch's active role 200
Kew student gardeners' visits to flower shows 19
Primula Conference (1913) 127, 133
Purdom's notebooks in archives i
Royal International Horticultural Exhibition (1912) 119–20
Wisley garden, Surrey 113, 200
see also The Garden (RHS magazine)
Royal International Horticultural Exhibition (1912) 119–20
rubber trees 7, 12
Russell, Bertrand 198
Russia, treaties with China (19th century) 54

Salomonsson, Stefan 88†
Sander, Frederick 12*
Sanderson camera ('Tropical' model) 46, 66*, 73, 76, 134, **205**
Sargent, Charles Sprague
  biographical details
    death 200
    portrait by John Singer Sargent 40
  Chinese Sargent–Veitch plant-collecting expedition (1909–12)
    agreement with Veitch 43–4
    appointment of Purdom 46
    criticism of Purdom's 1909 collecting results 75–6, 85–6
    criticism of Purdom's 1910 collecting results 89, 90–1
    criticism of Purdom's results compared to Wilson's 111, 113, 114–17
    criticism of Purdom's results to Meyer 121, 122, 123
    decision not to extend Purdom's contract 93, 94, 99–100, 103
    letter of introduction for Purdom 62
    memorandum of 'guidance' 47–8
    Purdom sending seeds of elms and *Caragana* 71
    Purdom sending statement of expenses to 73
    Purdom writing about 1910 expedition results 89, 90
    Purdom writing about Choni 98–9
    Purdom writing about Jehol 68
    Purdom writing about Kansu expedition 93–4
    Purdom writing about meeting with Veitch on return to London 111
    Purdom writing about Morrison's advice re. Shansi/Shensi provinces 77
    Purdom writing about travelling to Xi'an and sending seeds 83–5
    Purdom writing about trees 68, 71
    Purdom writing about Wutai-shan passport 71, 73
    Wilson's meeting with Purdom 58–9
    Wilson's reaction to Purdom's appointment 58
  horticulture
    Aesculus chinensis, interest in 89, **106**, 107–8, 111, 112
    Arnold Arboretum Directorship 14, 39, 75, 200
    Arnold Arboretum fundraising 115
    Crataegus, interest in 89†
    dessicationist theory, belief in 68‡
    Hers, herbarium specimens from 181–2
    Lilium regale, interest in 78–9, 89
    Meyer, dealings with 43, 48, 75
    peonies (herbaceous), interest in 70
    peonies (tree peonies), interest in 48, 83, 84*, 88, 91–2, 93
    Royal International Horticultural Exhibition, plants from seeds sent by 119
    Wilson 1910 expedition to China 59, 78–9, 81, 89, 90, 91
  Purdom
    information about to Farrer 127, 133
    plant material from (1920–21) 182–3
    rejection of Arnold Arboretum post by 39–41
    unfair criticism of 196
    see also Arnold Arboretum, Boston; Chinese 1909–12 Sargent–Veitch plant-collecting expedition
Sargent, John Singer, portrait of Charles Sprague Sargent **40**
Satanee *see* Sha-tan (Satanee) mountains
saxifrage 159
Scapanulus oweni (Kansu mole) 102
Scottish Enlightenment 19
sea buckthorns 84
Second Revolution (China, 1913) 126
self-improvement 4–5, 8, 129
  see also education
Senecio 126
Shaanxi *see* Shensi (now Shaanxi) province
Shadong 'concession' 180
Shanghai
  cosmopolitan city 61
  'Shanghailanders' 195
  treaty port 67
Shansi (now Shanxi) province 43, 48, 65, 76–7, 79, 83, 86
  deforestation **86**
Sha-tan (Satanee) mountains 136, 139
Shensi (now Shaanxi) province
  language 146
  Ordos desert 84, 109
  Purdom's 1910 expedition *see under* Chinese 1909–12 Sargent–Veitch plant-collecting expedition
Shensi Relief Column 101–2

INDEX

Sherfesee, William Forsythe 164, 165, 166, 167, 168, 177, 200
Shine, Edward 42
shrubs *see under* plants
Shulman, Nicola A. 230n129 (158)
Siberian pea tree (*Caragana*) 71
Sining (modern Xining) 99, 156, 157
Sitwell, Osbert 131
Smiles, Samuel 6, 8, 12
  Self-Help 4–5
Smith, Jack, Dr 93, 102
social mobility
  and education debate (1860s) 3–4, 5
  and Smiles' *Self-Help* book 4–5
  and working overseas 129
Soothill, Dorothea 107, **107**, 108
Soothill, Lucie (Mrs) 107, **107**, 108
Soothill, William 107
sorbus 79
Sowerby, Arthur de Carle 101–2
Stearn, William T. 129*
Strachey, Edward, Sir 38, 41–2
strikes
  Kew strike 39
  and Trade Disputes Act (1906) 39*
Stuart Low and Co. 12
  see also Hugh Low & Co nursery (Bush Hill, Enfield)
sumac trees 183
Summerbell, Thomas 38, 41, 42
Sun Yat Sen 101, 104, 134
swamp cypress (*Taxodium distichum*) 180–1, **209**
syphilis 15–16
Syringa (lilac) 84
  Syringa pubescens juliana 'Hers' 182
Szechuan Province 78, 79

Taff Vale Railway case 39*
Tai-Pei-Shan (now Taibai Shan, Qinling range) 87, 88, 90
Taiping Rebellion (China, 1850–64) 53, 54
takin (*Budorcas taxicolor*) 102
Tang dynasty 145
Taxodium distichum (American swamp cypress*)* 180–1, **209**
Tebbu people 97, 98
Temple of Heaven, Peking **107**, 108, 167
theodolites 65, 65*
Theophrastus 20
Thiselton-Dyer, William, Sir 17, 22–3, **22**, 26, 27–9, 33, 36, 59
Thundercrown *see* Leigu Shan mountain (Thundercrown)
Tian Wenlie 176*
Tibet
  expedition funded by Duke of Bedford 93
  Farrer–Purdom expedition (1914) 137–9
  Meyer's Chinese staff's refusal to go to 145
  political situation (first decade of 20th century) 96, 137–8
  staple food 64
  Tibetan peasant girl (Purdom's photograph) **76**
  see also Choni (now Joni or Zhuoni)
Tien Tang monastery 157, 159
The Times, Morrison as Peking correspondent 58, 76, 100†, 125
Tongxi Emperor *see* Tung-chih (Tongxi) Emperor
Trade Disputes Act (UK, 1906) 39*
trade unions
  Kew Employees Union 27–8, 30–1, 34, 36, 37, 39, 41, 46
  Taff Vale Railway case 39*
  Trade Disputes Act (1906) 39*
Trafalgar, Battle of, centenary of 30
trans-Siberian Express 79, 111, 133
treaty ports 54, 56, 67

tree peonies (Moutan peonies) 48, 83, 84, 88, 91–2, 93, 126, 136
  Paeonia rockii 92†
  Paeonia suffruticosa 92
  Paeonia veitchii var. *woodwardia* 92
trees
  on Brathay estates 3, 9–10
  Dallimore's specialist knowledge of 21
  Purdom's love for/specialist knowledge of 10, 21–2, 40
  specific species *see under* plants
  see also forestry; Forestry Adviser to Chinese Government
Trees of Britain and Ireland (Augustine Henry and Henry John Elwes) 46
Tsao Ju-lin (Cao Rulin) 180
Tseng Kuo-Fan (Zheng Guofang) 54
Tsin Shih Tse temple, Peking **106**
Tsingling mountains *see* Qinling mountains (Tsingling mountains, Shensi province)
Tso Zongtang (Zhuo Zongtang) 54
Tsu-An *see* Ci-An (Tsu-An)
Tsu-Hsi *see* Cixi (Tsu-Hsi), Empress Dowager
tulip fever (1630s) 11
Tung-chih (Tongxi) Emperor 51*, 53–4

Uganda, translocation of *Eucalyptus* species from Australia 7
Ulmus pumila 'pendula' (Peking elm) 112§, 193
United Kingdom (UK)
  domestic issues
    caste system 129
    Corn Laws, repeal of (1846) 29
    Education Acts (England and Wales, 1870 and 1880) 4
    General Election (1906) 33–4
    snobbery 183
    suffrage before WWI 156

249

INDEX

Trade Disputes Act (1906) 39*
Easter Rising (Dublin, 2016) 168
foreign policy
  Burma, control over 55
  Canton, battle of (1857–58) 53
  China, opium agreement with (1909) 85
  China, Opium War with 52–3
  China, treaties with (19th century) 54
  Old Summer Palace (Peking), burning of 53
  Tibet as buffer between India and Russia 96
WWI
  1916 year 168–9
  Chinese Labour Corps (CLC) 172–3
  'Derby scheme' and conscription (1915) 155–6
  Foreign Office propaganda unit 172†
  Military Service Act (1916) 171

see also British Army in India; British Empire; British Legation, Peking; British secret services

United States, treaties with China (19th century) 54
University of California Davis Library, Frank Meyer collection iv
Unwin, A. H. 21
Utilitarianism 19

Veitch *see* Coombe Wood nursery; James Veitch & Sons; *separate family members*
Veitch, Harry
  biographical details
    death 200
    James Veitch & Sons, running of 14–16
    knighthood 119
    portrait **14**
    retirement 127
    RHS Primula Conference, attending 127

Royal Horticultural Society, active role in 200
Royal International Horticultural Exhibition, organisation of 119–20
selling of *Paeonia Veitchii* to Woodward 92
Chinese Sargent–Veitch plant-collecting expedition (1909–12)
  asking Purdom to join expedition 43–5, 58, 63
  asking Sargent to be indulgent towards Purdom 91
  concerns about Purdom's welfare 104–5
  decision not to extend Purdom's contract 93, 94, 99–100, 103
  disappointed with overall results 111
  expectations from expedition 48
  happy with Purdom's first year's collecting 75
  Henry's advice 46
  "If the trees and plants were not there, he could not send them" 94, 111, 116
  letters of introduction for Purdom 47, 59, 62
  meeting with Purdom back in London 111
  on Purdom being a poor correspondent 117
  on Purdom "throwing himself into the work" 47
  Purdom writing about ambush 106, 107
  Purdom writing about anemones and sending roots 68, 70–1
  Purdom writing about Choni 98
  Purdom writing about Dolonor 74
  Purdom writing about ill-health 70
  Purdom writing about Kansu expedition 93, 94
  Purdom writing about preparations for 1910 season 77–8

Purdom writing about Qinling and hostile pilgrims 87–8
Purdom writing about Sanderson camera problems 73
Purdom writing about travelling to Xi'an and sending seeds 83–5
Purdom writing about wild animals and highwaymen 74
Purdom after the expedition
  claiming not to have his address 121–2, 122*, 123
  failure to recognise his efforts 119–20
  last dealings 120
  not helping him with future employment 123
  'Seeds collected by W Purdom' register to Kew 112–14
  seeds from Purdom via Farrer (1916) 183
see also Chinese 1909–12 Sargent–Veitch plant-collecting expedition; Coombe Wood nursery; James Veitch & Sons
Veitch, James 13–14
Veitch, James Herbert 14–16
  Hortus Veitchii 15*
  syphilis 15–16
Veitch, James junior 14
Veitch, John 13
Veitch, John Gould 14, 100
Veitch, John Gould junior 15
Veitch, Robert 14
Verganni family 175
Versailles Peace Conference (1919) 180
viburnums 84
  Viburnum fragrans (later *V. farreri*) 136
Victorian Gothic Revival style 3*
Vietnam *see* Annam kingdoms (now Vietnam)

Wallace, Harol 102–3, **103**
Wallis, E. J. 46
Wang Ching Ch'un (Wang Jing Chun) 177, 201–2, 218

250

INDEX

Warley Place, Essex 119
warlords (in China) 54
  warlord era (1910s–20s) 176
Waterhouse, Alfred 3*
Watson, William 17–18, 19, 25, 26–7, 28–9, 36, 47
Watt, Alistair 219
Weichang district, Chihli (Zhili) province 47–8, 71, 73, 74, 85–6, 109
Weichang Hunting Forest 47, 68, 75, 77, 94
Wellingtonia gigantea 9, 14*
Whampoa, Treaty of (1844) 52
White (Low's collector) 12*
White Craggs (nr Ambleside) 124, **124**, 127, 197
White Wolf *see* Bai Lang ('White Wolf')
wildlife 74, 84, 102
Willmott, Ellen 119
willows 68, 199
Wilson, Ernest Henry
  biographical details
    best-selling books 115
    death in car accident 200
    lionised by American press/public 200
    self-aggrandisement 74
    work (taxonomical) at Arnold Arboretum 78
    work as Keeper of Arnold Arboretum 200
    work at Kew 59
  on China
    after Empress Cixi's death 58
    south-western China as "botanical paradise" 81
    "Wanderings in China" lecture (Kew) 59
  Chinese expedition (1910–11)
    different goals/areas compared with Purdom's expedition 78–9, 81
    hiring same staff 59
    leg injury 89, 90, 91, 115
    plant material unfairly prioritised by Coombe Wood 123
    records of specimens at Arnold Arboretum 113
    Sargent's comparison with Purdom's expedition 111, 113, 114–17
  horticulture and plant hunting
    Chinese expeditions (pre-1910) 43, 45, 46, 59, 75, 78, 114
    glass plates, sending to London for processing 46*
    Lilium regale bulb shipment disaster 78
    New Hardy Plants from Western China (introduced through Mr. Ernest Wilson) 119
    Plantae Wilsonianae 85*, 92, 112
    R. Moyessi and *R. Willmottiae* 120*
    Royal International Horticultural Exhibition, plants introduced by at 119–20
    surplus seeds dumped by Coombe Wood 114
    "Wilsonian weeds" 114*
  James Herbert Veitch, criticism of in *Aristocrats of the Garden* 15†
  Meyer
    briefed by 59
    unfairly overshadowed by 116
  Purdom
    briefed by 47, 58–9
    unfairly overshadowed by ii, 115–16
Wisley, Surrey, Royal Horticultural Society (RHS) garden 113, 200
Wolvesden House ('Wolf stone' valley, China) 157–8, **157**, **158**, 159, 160
Woodward, Robert 120, 151–2
Paeonia veitchii var. *woodwardia* 92
World War I 140, 151–2, 155–6, 161, 168–9, 170–1, 172–4, 183
  see also under United Kingdom (UK)
Wu, General 149
Wu Ting-Fan, Dr (Ng Choy or Ng Achoy) 125, 228n57 (67)
Wutai-shan 68, 71, 73, 75
  camp 70

Xi'an 48, 77, 83–4, 85, 93, 136, 164
  Xi'an massacre (1911) 100, 101–2, 103
Xianfeng, Emperor of China 53, 54
Xinhai (Republican) Revolution (China, 1911) 100–2, 103–6, 116, 125–6
Xining *see* Sining (modern Xining)
Xinyang *see* Purdom Forest Park, Xinyang (formerly Mount Jigong forest station)

Yakub Beg 54
Yang-Chi-Ching, Prince of Choni 94, 95, **95**, 96–8, 200
Yangtze Gorges 160
Younghusband, Francis 96
Yuan Shih K'ai, General (later President)
  death 167–8
  Morrison working on behalf of 125, 198
  proclamation of himself as Emperor 161*, 163, 163*
  role in Xinhai (Republican) Revolution 101, 102, 103–4
  threats, murder and bribery 126
  Viceroy of Koko-nor, arrested by 160
  'White Wolf' (Bai Lang)'s army against 134

Zhang Bing Hua, Viceroy of Koko-nor **148**, 149, 157, 160
Zhang Jian *see* Chang Chien (Zhang Jian)
Zheng Guofang *see* Tseng Kuo-Fan (Zheng Guofang)
Zhili *see* Chihli province (modern Zhili); Weichang district, Chihli (Zhili) province
Zhou Xuexi *see* Chou Hsuëh-Hsi (Zhou Xuexi)
Zhou Ziqi *see* Chow Tzu-Chi (Zhou Ziqi)
Zhuo Zongtang *see* Tso Zongtang (Zhuo Zongtang)
Zhuoni *see* Choni (now Joni or Zhuoni)
Zizania latifolia (Manchurian water rice) 111

Francois Gordon

Francois Gordon retired from the Foreign Office in 2009 after 30 years mostly spent in Africa. He has always sought relief from the intellectual demands of political reporting and analysis in gardening and is a lifelong student of the history of plants and gardens.

Francois first encountered Will Purdom in 1986, and decided that he had not received the attention or the credit he deserves. Although uneasily aware that Will, a very private man, might not have welcomed the attention, he has spent five years searching archives on four continents for insights into Will's life and work and hopes he may have done justice to a fine and honourable man.

Francois lives and gardens with his wife Elaine in Kent. This is his first book.